Caves of Knoxville and the Great Smoky Mountains

Caves of Knoxville and the Great Smoky Mountains

Larry E. Matthews

National Speleological Society

Library of Congress Control Number
ISBN 978-1-879961-30-2

First edition June 2008

Produced by the
National Speleological Society
Special Publications Commiteee
G. Thomas Rea, Chairman

Front Cover photograph by Kenneth Storey: Cheryl Pratt stands next to massive flowstone formations in Gap Cave, April 2005.

Back Cover photograph by Bob Biddix: Erica Sughrue by a rimstone dam and flowstone in Morrill Cave, March 22, 2008.

Book design by
Greyhound Press
Cloverdale, Indiana

GREYHOUND PRESS

Published by
The National Speleological Society
2813 Cave Avenue
Huntsville, Alabama 35810-4431
USA

Dedicated to

Thomas C. Barr Jr

Thomas C. Barr Jr, has been an inspiration to Tennessee cavers for 47 years, ever since his monumental book, ***Caves of Tennessee***, was published in 1961. This remarkable book, which describes over 700 Tennessee caves and is 567 pages long, has been an inspiration to three generations of cavers. Unlike any bulletin previously published by the Tennessee Division of Geology, *Caves of Tennessee* soon sold out and had to be reprinted. It has since been reprinted numerous times and is still in print today. Even today, it is one of the best-selling books sold by the Tennessee Division of Geology.

Other Books by This Author

If you enjoyed this book, you may wish to read other books by this author. Listed below are other titles about caves, cave exploration, and geology.

Bibliography of Tennessee Speleology (4th Edition), published by the Tennessee Cave Survey in 1994. This massive, 420-page-long volume is the definitive source of information on publications concerning Tennessee caves from 1806 to 1994. It cross-references 2,740 books and articles with 2,427 cave names, and 557 different subjects.

Bibliography of Tennessee Speleology – Update to 4th Edition, published by the Tennessee Cave Survey in 2001. This 73-page-long volume adds another 271 books and articles that are cross-referenced with 373 cave names and 234 different subjects.

Descriptions of Tennessee Caves, published by the Tennessee Division of Geology, 1971, 150 pages. This book provides descriptions of 316 caves, along with numerous maps and photographs. There is a special section on the history of saltpeter mining in Tennessee during both the War of 1812 and the Civil War.

Springs of East Tennessee, published by the Tennessee Cave Survey in 2003, 110 + viii pages. Locations and basic information on 766 springs in east Tennessee.

Springs of Middle Tennessee, published by the Tennessee Cave Survey in 1997, 120 + iv pages. Locations and basic information on 795 springs in middle Tennessee.

Springs of West Tennessee, published by the Tennessee Cave Survey in 2004, 45 + vii pages. Locations and basic information on 59 springs in west Tennessee.

Dunbar Cave, The Showplace of the South, published by the National Speleological Society, 2005, 152 pages. This is the fascinating story of how this cave, first explored by Native Americans 8,000 years ago, grew to become one of America's major caves. Includes the complete story, maps, and photographs as modern explorers add miles of passage to this former show cave that had been explored for the last 150 years.

Cumberland Caverns (2nd Edition), published by Greyhound Press, 2005, 188 pages. The story of how this cave, first discovered by Aaron Higgenbotham in 1810, has grown to become one of the world's largest and most famous caverns, including historical anecdotes from the last century and the latest breakthroughs, including the 1976 discovery.

Big Bone Cave, published by the National Speleological Society, 2006, 224 pages. The fascinating story of America's largest saltpeter mine, located 50 miles north of Chattanooga. Also the site of the discovery of giant ground sloth skeletons, an ice-age jaguar skeleton, and several American Indian mummies.

Caves of Chattanooga, published by the National Speleological Society, 2007, 192 pages. The exploration, use, and commercial development of nine Chattanooga-area caves. Learn how for 8,000 years, the caves were used by the Cherokees until their relocation in 1838; how settlers explored and used these caves; how the caves were mined for saltpeter, the main ingredient of gunpowder, during both the War of 1812 and the Civil War. Read never-before-published stories of fascinating discoveries deep inside Lookout Mountain. Be amazed by the Great Cave Hoax of 1927.

Afterthoughts

Francis McKinney
1973

As I trudge down the old road
With clothes torn, wet, and muddy,
A hawk wings down the slope across the valley
And trees ripple wildly in the wind.

I remember crawling slowly
In the water on the gravel
And squeezing through projections sharp and painful.

I remember climbing quickly
Up the ladder in the shower
And shivering as we mapped the last ten stations.

I remember looking closely
At the beauty in the darkness
And wondering when we'd get back there again.

Let the hawk have his sunlight!
While I pursue my pleasure—crawling, climbing, searching—
I match his joy measure for full measure.

Table of Contents

Caves of Knoxville and the Great Smoky Mountains

Acknowledgements

Persons who have assisted in providing information for this book and in reviewing the various chapters are listed below alphabetically. Their assistance is sincerely appreciated. I hope no one was left out.

Caver **Laurie Adams** published a booklet on Morrill Cave in 1984 that was a valuable source of information on that cave. He also proof-read the chapter on Morrill Cave.

Archaeologist **Donald B. Ball** provided fascinating information on Alum Cave and proofread that chapter.

Cave biologist **Thomas C. Barr Jr's** classic book, *Caves of Tennessee*, provided the logical starting point for information on ten caves described in this book that are located in Tennessee. Tom answered many questions for me during the preparation of this book.

Cave photographer **Bob Biddix** provided outstanding photographs for several chapters in this book. I can always count on Bob to help.

Caver **Jeff Bowers** provided exciting information on the exploration and mapping of Gap Cave during the 1980s and 1990s.

Caver **Anne Elmore** provided valuable information on Gap Cave.

Caver **Sandra S. Sherlin** provided photographs of Morrill Cave and interesting historical information about the cave.

Professional geologist and spelean historian **Angelo I. George** provided valuable information on the early history of saltpeter mining in the Cumberland Gap area. He was also kind enough to proofread several chapters of this book and provide helpful suggestions.

Librarian **Annette F. Hartigan** at Great Smoky Mountains National Park provided access to Park records and provided other valuable information. She was also kind enough to proofread the chapters on Alum Cave and Gregorys Cave.

Roger and Susan Hartley, the owners of Appalachian Caverns, provided access to their cave and helped in every way possible.

Cave surveyor **Dan Henry** provided the recent map of Gregorys Cave.

As usual, my wife, **Betty Matthews**, was kind enough to put up with another one of my cave projects and accompanied me on several of the field trips to gather information and take photographs.

Biologist **Jonathan Mays** provided a copy of his Master's Thesis on Gregorys Cave and additional information on the cave.

Lost Sea manager **Lisa McClung** provided access to the cave, historical photographs, and reviewed the material in that chapter.

Caver **Gerald Moni** with the Tennessee Cave Survey answered numerous questions and searched the Survey files for information.

Caver **Cheryl Pratt** provided excellent photographs of English Cave and Gap Cave and also provided valuable information on these caves.

NSS Special Publications Chairman **Tom Rea** provided tremendous assistance in preparing this book for publication. Tom's contributions to caving and the NSS are truly inspirational. Certainly, Tom has been my mentor these past four years. Thanks for all you help.

Cave historian **Gordon L. Smith Jr.** provided access to his remarkable collection of commercial cave memorabilia.

Cave historian **Marion O. Smith** provided valuable information on saltpeter mining in east Tennessee and New Mammoth Cave.

Caves of Knoxville and the Great Smoky Mountains

Cave historian **Gary K. Soule** has a wealth of knowledge on the commercial caves of the United States. He was very generous in sharing material from his collection for use in this book, including brochures and postcards.

Cave photographer **Ken Storey** provided excellent photographs of Gap Cave.

Cave photographer **Lisa Storey** also provided excellent photographs of Gap Cave.

Cave historian **Jim Whidby** provided much valuable information on Cherokee Caverns, Gregorys Cave, Indian Cave, and New Mammoth Cave. He was very generous in sharing material from his collection for use in this book.

Numerous other persons, whose names do not appear above, were kind enough to answer questions, respond to e-mails, provide information, and offer advice. Their help was sincerely appreciated.

Illustrations

Caves of Knoxville and the Great Smoky Mountains

Introduction

With over 10 million visitors coming every year, Great Smoky Mountains National Park is the most visited National Park in the United States. Although there are no commercial caves inside the park boundaries that are open to tourists at this time, there are a number of commercial caves in the greater Knoxville and Great Smoky Mountains area that are open to the public. Many of the same tourists who visit the national park also stop and visit these commercial caves. Because tourists arrive and leave the park from all directions, commercial caves south of the park (The Lost Sea), west of the park (Cherokee Caverns, Indian Cave, and Tuckaleechee Caverns), north of the park (Forbidden Caverns and Gap Cave), and northeast of the park (Appalachian Caverns and Bristol Caverns) are all included in this book. All of these caves are in Tennessee with the exception of Gap Cave, which is just across the state line in Virginia. There are no commercial caves in the Smoky Mountains area in North Carolina, because the geology there is not suitable for the formation of large caves.

In addition to the eight commercial caves listed above, this book also documents the long and fascinating history of five other caves in this area that have been commercial caves in the past. Also included in this book is Alum Cave, a large rock shelter located in the Great Smoky Mountains National Park. Although not a true cave, Alum Cave has a long and interesting history and is visited by thousands of visitors to the park each year.

There are easily over 1,000 known caves in the area covered by this book. However, for the most part these are wild caves and on private property and they will not be the subject of this book. Persons interested in exploring wild caves will be given information on how to pursue this subject in Appendix A.

It is the author's wish that this guide to the commercial caves of Knoxville and the Great Smoky Mountains will greatly enhance the readers' visits to these caves by going into details of their history that a regular tour guide could not hope to convey in the short amount of time that a tour lasts. For those commercial caves that are no longer open to the public, this book attempts to preserve their rich history.

The book *Caves of Chattanooga*, which was published in April 2007, covers the area to the southwest of the area that is the focus of this book. Persons who enjoy this book will want to read *Caves of Chattanooga* and learn about the caves near Chattanooga that are located in portions of Tennessee, Alabama, and Georgia.

Chapter 1

Cherokee Caverns

The Discovery of Cherokee Caverns

American Indians explored Cherokee Caverns, as evidenced by cane torch stoke marks left on the walls. These early explorers would tap their torches against the wall to knock off embers and improve the flame. This activity left a very distinctive mark on cave walls and scattered fragments of charcoal on the floor.

Caucasians began to settle in east Tennessee in the second half of the 1700s. This settlement increased dramatically after 1775, when Daniel Boone led settlers along the Wilderness Road through the Cumberland Gap. This is the point where Kentucky, Tennessee, and Virginia meet at a natural gap through the Appalachian Mountains. Knox County was formed in 1792 and Tennessee became the 16th State in 1796.

The earliest record that we have of Caucasians exploring the cave is in about 1854, when a local farmer, Robert Crudgington, (1830–1913) discovered the cave. According to legend, Crudgington was hunting on the hillside when he noticed fog emerging around rocks. After moving the rocks, he entered the caverns and at the time was thought to have been the first person to see them. The faint traces of earlier Indian exploration were not noticed at that time. The Indians may have used an entrance that no longer exists. Crudgington bought 800 acres of farmland, including the entrance to the cave, in 1866. It is not known what, if any, name was used for the cave during this time period, but in all likelihood it would have been referred to locally as Crudgington Cave.

History of the Cave

The first owner of the cave was Robert Crudgington, who purchased 800 acres of land, including the entrance to the cave, in 1866. The cave was first commercialized in 1929, by Crudgington's daughter, Margaret Crudgington Gentry, under the name **Gentrys Cave** and in 1930, the cave was renamed **Grand Caverns.** The caverns were well advertised and were visited by many people. Margaret Crudgington owned the cave until her death in 1946. In 1947 the family sold the cave and it was leased and reopened under the name *Atomic Caverns*. Since nearby Oak Ridge, Tennessee, was the secret city where the atomic bomb was developed during World War II (1941–1945), it would seem reasonable that the name Atomic Caverns was taken from the cave's proximity to the Atomic City, but Jim Whidby has been told that the cave was re-named for a large flowstone formation in the cave that was thought to resemble the mushroom cloud of an atomic bomb explosion at Bikini Atoll.

Cherokee Caverns manager Jim Whidby provided a handbill advertising a radio show to be broadcast live from Atomic Caverns on Sunday, July 16, 1948. The featured performer for this show was Homer Harris, known as the "7 Foot Smilin' Cowboy," along with "His Educated Trick Horse Stardust." The show was a production of WIBK Radio in Knoxville, Tennessee. Jim Whidby also provided an 8 by 10 photo of Homer Harris, his horse, and his band taken inside the cave. It is interesting to note that one of the musicians has no arms! He is supposed to have played the fiddle with his feet. One performance was given in the Crystal Ballroom of the caverns.

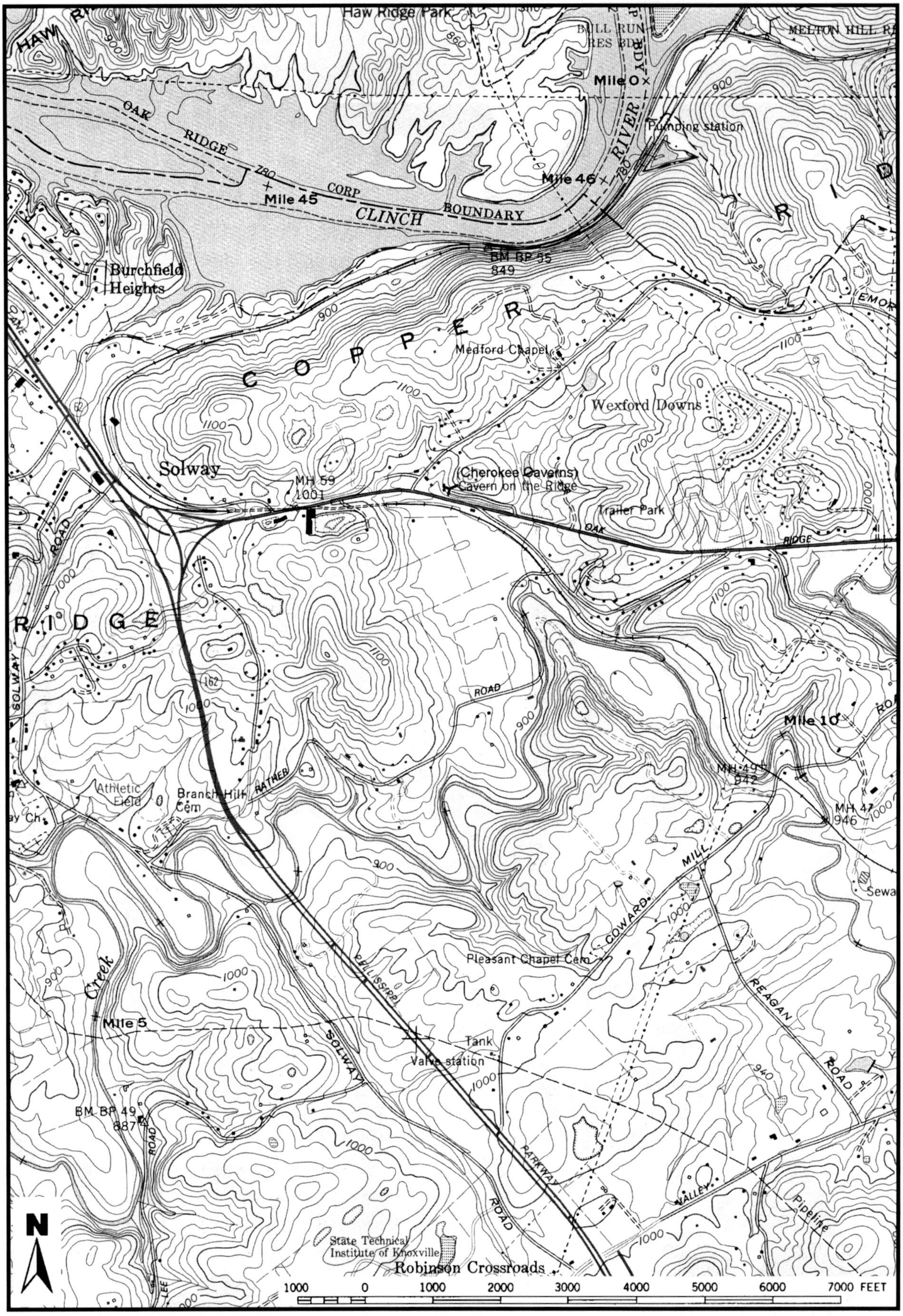

Figure 1.1. Topographic map showing the location of the entrance to Cherokee Caverns.

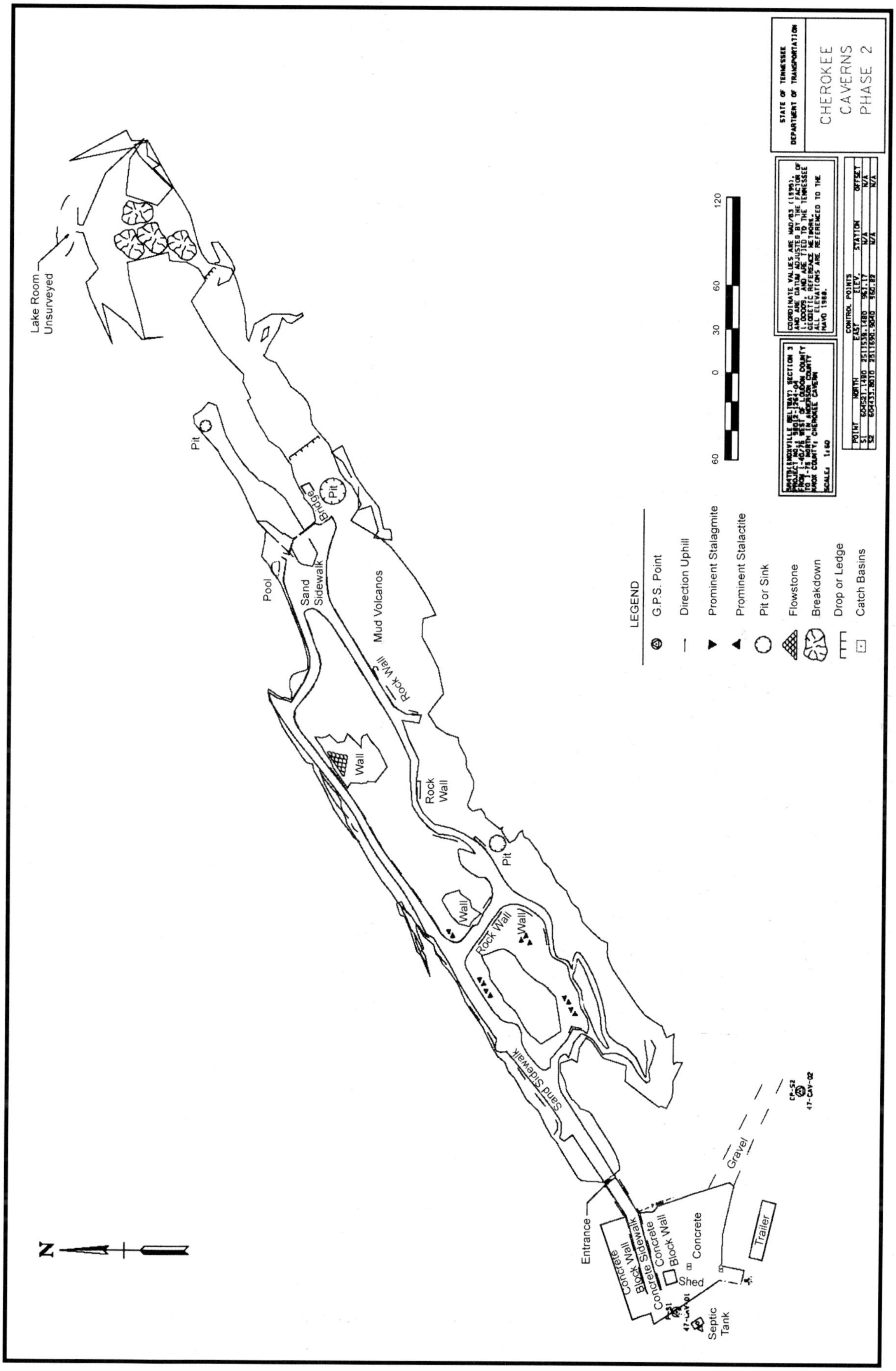

Figure 1.2. Map of Cherokee Caverns prepared by the Tennessee Department of Transportation.

Figure 1.3. Tombstone for Robert Crudgington (July 3, 1830–August 10, 1913). Photo by Jim Whidby.

A restaurant was built above the cave entrance in the 1960s, the trail and the lighting in the cave were improved, and the cave was renamed **Caverns of the Ridge**. More name changes followed. In 1970, the cave was renamed **Cherokee Firesite Ceremonial Caverns**. A few years later, this was shortened to **Cherokee Caverns**, the name still in use today. On October 1, 1980, a fire destroyed the restaurant built over the entrance to the caverns and the cave was closed to the public. Unfortunately, for the next nine years the cave sat closed and unguarded and vandals broke into the cave on a regular basis and did significant damage.

In 1989 Scott and Janet Dixon moved to the region and began to redevelop the cave as a commercial attraction. In October 1989 they held the first Haunted Cave, an event that still continues today. Cherokee Caverns was designated a Tennessee Natural Area on November 14, 1989. In January 1992 the Dixons moved away. Jim Whidby now manages the cave and the 20th Annual Haunted Cave will be held in the fall of 2008.

What's In A Name?

Cherokee Caverns seems to have the unusual distinction of being shown commercially under more names than any other Tennessee cave. Below is a list of the names that have been used, and the dates that they were first used.

Year	Name
1929	Gentry's Cave
1930	Grand Caverns
1947	Atomic Caverns
1950s	Caveman's Palace
1950s	Palace Caverns
1960s	Caverns of the Ridge
1970	Cherokee Firesite Ceremonial Caverns
1971	Cherokee Caverns

The 1990 photorevised version of the Lovell, Tennessee, 7.5-minute series topographic Map (138-NW) shows the location of the entrance to the cave along with the name Cavern on the Ridge. Apparently, the person revising the original 1968 version of the map misinterpreted the name of the cave that was in use in the 1960s. The name Cavern on the Ridge was never used as an official commercial name of the cave, yet ironically, it is the only name to ever appear on a map.

A Description of the Cave

Cherokee Caverns is open to the public at the present time, but on a limited basis. The book, *Descriptions of Tennessee Caves*, which was written by Larry E. Matthews in 1971, gives the following written description of the cave:

> Grand Caverns has been known by several different names during its history as a commercial cave. The name in use in late 1968 is Atomic Caverns. The original entrance was 5 feet in diameter and sloped downward into the cave. The entrance has been enlarged artificially to 15 feet in diameter.

Figure 1.4. Erica Sughrue at the gate to Cherokee Caverns. Photo by Bob Biddix, May 18, 2008.

The portion of the cave open to the public consists of a very large and profusely decorated Allegheny-type passage. The passage averages 80 feet wide and 20 feet high for 500 feet. Past this point it is not commercially developed, and its extent is not known to the writer. The formations are so numerous in places that they form walls which compartment the main passage into sections.[1]

As it turns out, the cave extends only a few hundred feet past the commercial trails. Refer to Figure 1.2 for an accurate map of the cave.

Several maps of Cherokee Caverns have been made over the years. The most recent map was made by the Tennessee Department of Transportation as a part of their study to build a new road in the vicinity. This map is shown as Figure 1.2. As a result of this survey, it was determined that if the new road was built as proposed, the resulting road cut would intersect Cherokee Caverns at the northeast end of the cave, thereby destroying part of this historic cavern.

The official Tennessee Cave Survey length for Cherokee Caverns is 1,500 feet. This would compare favorably with the Tennessee Department of Transportation map.

Cherokee Caverns Today

Cherokee Caverns is open to the public at the present time, but on a limited basis. The cave is open for ten days before Halloween and a Haunted Cave Tour is given at this time. Tours can also be arranged in advance for groups such as scout groups, school groups, or any other group wishing to book a guided tour of the cave. However, there is no visitor center

1 Larry E. Matthews, *Descriptions of Tennessee Caves*, Bulletin 69, Tennessee Division of Geology, 1971, pp 55–58.

Figure 1.5. Old postcard of "An Underground Fairyland – Grand Caverns, Knoxville, Tennessee." From the Speleo Archives of Gary K. Soule.

Figure 1.6. Old postcard of "The Caverns With A Million Stalactites." From the Speleo Archives of Gary K. Soule.

Figure 1.7. Erica Sughrue on the tourist trail in Cherokee Caverns. Photo by Bob Biddix, May 18, 2008.

Figure 1.8. Erica Sughrue in an area where the tourist trail has been excavated to walking height. Photo by Bob Biddix, May 18, 2008.

Figure 1.9. Old postcard of "One of the Most Spectacular Caverns in America – Grand Caverns, Knoxville, Tenn." From the Speleo Archives of Gary K. Soule.

Figure 1.10. Old postcard labeled on the back, "Fallen Stalagmite with Other Stalagmites That Have Formed on it." From the Speleo Archives of Gary K. Soule.

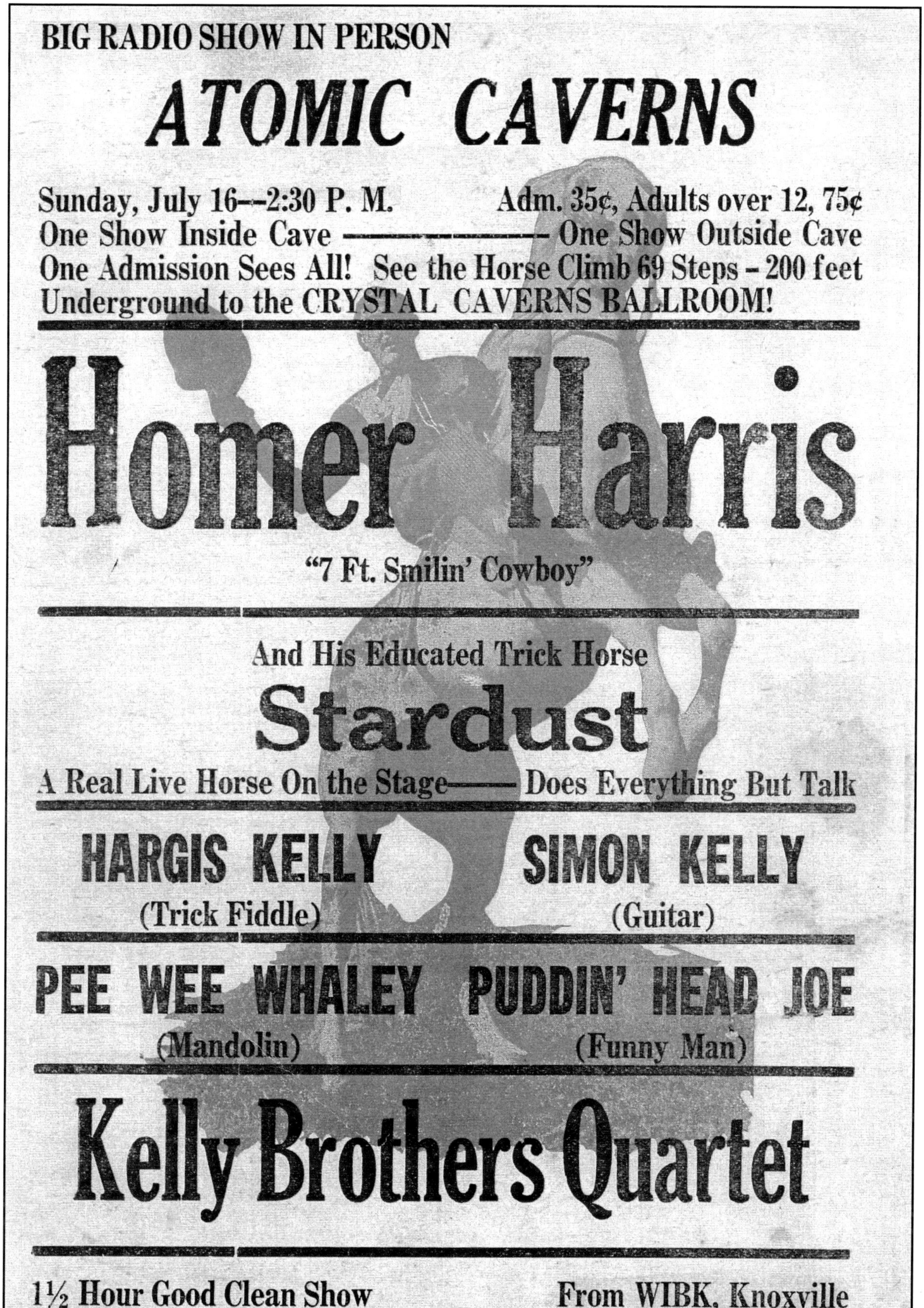

Figure 1.11. A 1948 advertisement for Atomic Caverns, promoting a radio show to be broadcast live from the cave. From the Jim Whidby collection.

Figure 1.12. Photo of Homer Harris and his horse, Stardust, inside Atomic Caverns in 1948. From the Jim Whidby collection.

Figure 1.13. Photograph of performers in the Atomic Caverns' Crystal Ballroom in 1948. From the Jim Whidby collection.

Figure 1.14. Erica Sughrue among beautiful flowstone formations in Cherokee Caverns. Photo by Bob Biddix, May 18, 2008.

Figure 1.15. Erica Sughrue beneath hundreds of stalactites on the ceiling of Cherokee Caverns. Photo by Bob Biddix, May 18, 2008.

Figure 1.16. Erica Sughrue beneath massive stalactites in Cherokee Caverns. Photo by Bob Biddix, May 18, 2008.

Figure 1.17. Erica Sughrue by a guardrail in profusely decorated Cherokee Caverns. Photo by Bob Biddix, May 18, 2008.

Figure 1.18. Old postcard labeled on the back, "Colorful Stalagmites and Stalactites Join to Form Columns of Incredible Beauty." From the Speleo Archives of Gary K. Soule.

at the present time and there are no permanent restrooms. Portable toilets are brought to the site for the Haunted Cave Tours and tents are used to house the food sales and other activities.

The road to the cave is gated at the highway and the cave entrance is also securely gated. The cave is electrically lighted and the trails are handicap accessible. For information on tours, you may contact the cave management at:

cherokeecaverns@aol.com

Guided tours may be scheduled for groups

Figure 1.19. Photo of the restaurant and ticket office before it burned in 1980. From the Jim Whidby collection.

Figure 1.20. Front of an old brochure for "Tennessee's Caverns of the Ridge." From the Speleo Archives of Gary K. Soule.

Figure 1.21. Front of an old brochure for "Caverns of the Ridge" showing an Indian Head logo. From the Speleo Archives of Gary K. Soule.

of 12, or more. The cave is also open at least once during the summer for a chance to cool down in the cave and once during the winter for a chance to warm up in the cave. These events are advertised in the Knoxville and Oak Ridge area.

Directions

From Knoxville, drive west on Interstate 40 for 14 miles to Exit 373. Go north on Highway 162 for 6.1 miles. Turn right onto Highway 62 (the Oak Ridge Highway) and go 0.8 miles. The driveway to Cherokee Caverns is on the left.

Tennessee
REGISTER
of
NATURAL AREAS

This is to certify that the
Tennessee Department of Conservation
has found **Cherokee Caverns**
to include highly significant natural features according to the Natural Areas Preservation Act TCA 11-1712. As a result of this finding and for the purposes of recognition and protection, this area has been officially included on the Register of Natural Areas.

Date 11-14-89

Commissioner of the Department of Conservation

Figure 1.22. Certificate designating Cherokee Caverns a Tennessee State Natural Area on November 14, 1989. Courtesy of Jim Whidby.

Figure 1.23. Cherokee Caverns "Ghost Host" button, 9th Annual (Haunted Cave) 1997. From the Speleo Archives of Gary K. Soule.

Figure 1.24. Discount coupon for 2006. From the Jim Whidby collection.

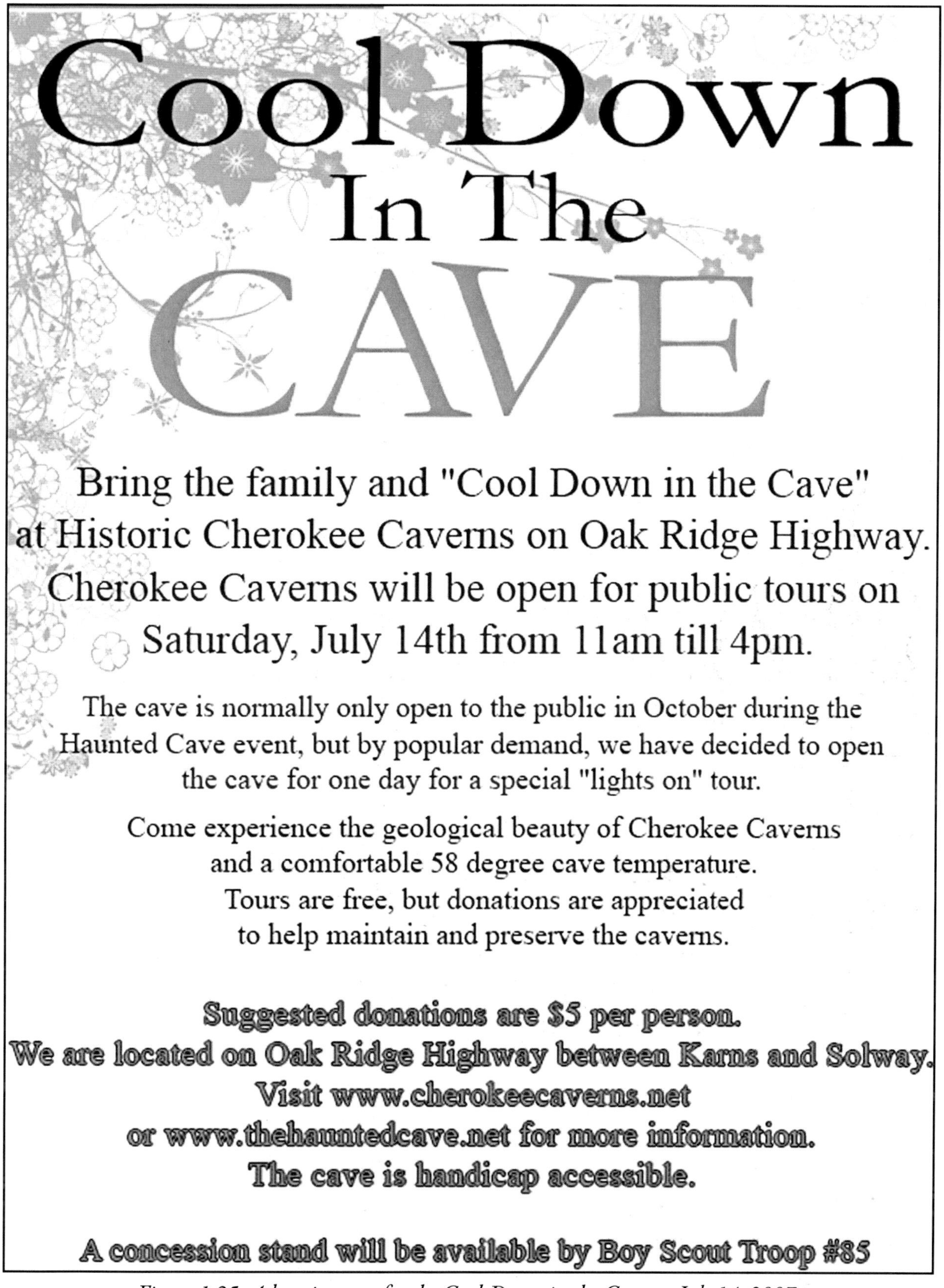

Figure 1.25. Advertisement for the Cool-Down in the Cave on July 14, 2007. From the Jim Whidby collection.

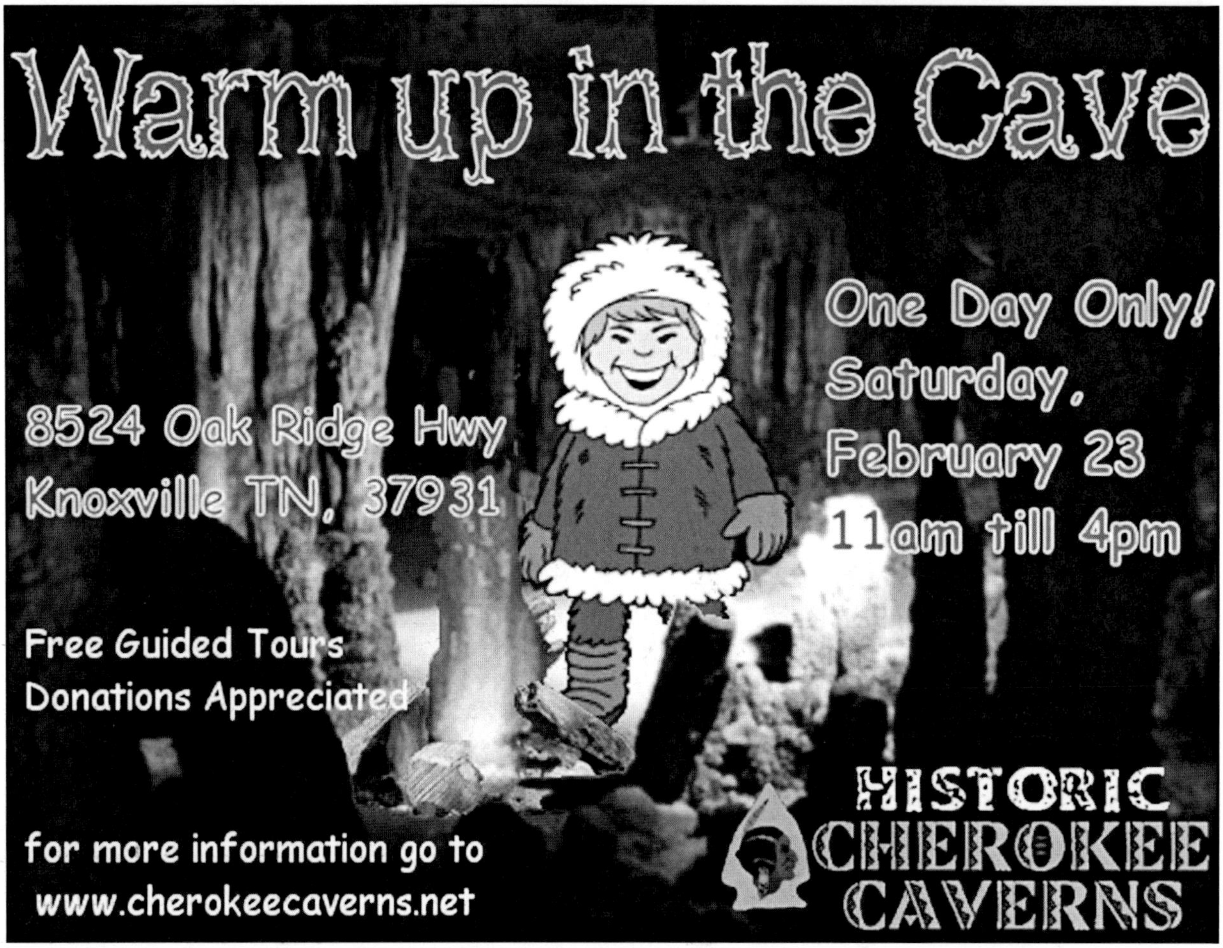

Figure 1.26. Advertisement for the Warm-Up in the Cave on February 23, 2008. From the Jim Whidby collection.

Figure 1.27. Erica Sughrue on the tourist trail in Cherokee Caverns. Photo by Bob Biddix, May 18, 2008.

Figure 1.28. Front of an old brochure for "Cherokee Fire Site Ceremonial Caverns." From the Speleo Archives of Gary K. Soule.

The address for the cave property is 8524 Oak Ridge Highway. The driveway to the property is gated.

Side Trips

Nearby Oak Ridge is home to the American Museum of Science and Energy. This museum opened in 1949 in an old wartime cafeteria, and was originally named the American Museum of Atomic Energy. Its purpose was to educate the public on the many peaceful uses of atomic energy. In the early days, you could place a dime in a machine and it would be subjected to radiation and come out in a souvenir package. If you held the dime close to one of the nearby Geiger counters, an excited "beep-beep-beep" would assure you that the dime was indeed radioactive. After this, of course, the child would place the treasured souvenir dime back in his pocket and subject his testicles to the radiation for the remainder of the day. The museum was moved to its current location in 1975 and was renamed the American Museum of Science and Energy in 1978. The museum is open daily, except for Christmas, Thanksgiving, and New Year's Day.

When you leave Cherokee Caverns, travel west on Highway 62, the Oak Ridge Highway, for 6.4 miles. This will take you into the city of Oak Ridge, which was built early in World War II for the purpose of conducting the secret research which resulted in the development of the world's first atomic bomb. After you enter Oak Ridge, continue through three traffic lights and turn right onto Tulane Avenue. The museum is a short distance ahead and on the left. The street address is 300 South Tulane Avenue and the telephone number is 865-576-3200.

Figure 1.29. Erica Sughrue among beautiful flowstone formations in Cherokee Caverns.
Photo by Bob Biddix, May 18, 2008.

Chapter 2

English Cave

The Discovery of English Cave

In all likelihood, English Cave was discovered soon after settlers moved into this area. It is five miles southeast of Cumberland Gap and close to the Powell River. However, no record of its initial discovery has been located.

Dr Thomas Walker located and named Cumberland Gap on April 13, 1750. Twenty-five years later, Daniel Boone followed the Wilderness Road through Cumberland Gap and into Tennessee and Kentucky, so settlers were moving into this area in significant numbers after 1775, and these settlers probably discovered English Cave sometime before 1800. Tennessee was already a state by 1796 and Claiborne County, where English Cave is located, was created in 1801.

The Four Seasons Hotel

Perhaps the earliest commercial use of English Cave occurred when the nearby Four Seasons Hotel offered buggy rides to the cave and conducted tours. The Four Seasons Hotel also developed Cumberland Gap Cave as a commercial cave. Cumberland Gap Cave is described in Chapter 3 under its current name, Gap Cave.

Barr's (1961) spelean history research reveals: "In the early part of the 20th century the cave was developed commercially by a utilitarian socialist colony of Englishmen at Cumberland Gap, and it still bears the name "English Cave." The rotting remains of wooden stairways, walkways, and bridges may be seen in many places."

An English land company, the American Association Limited, built the Four Seasons Hotel in 1892. The chief agent for this company was Colonel Alexander Alan Arthur. At the time, it was reported by its promoters to be the largest hotel in the United States. The building was four stories high with a frontage of 700 feet, a lobby 75 feet square, and a dining room 50 feet by 160 feet. It was reported to contain 700 rooms. A railroad spur from nearby Arthur, Tennessee, was built to provide access for private railroad cars. Also included in the complex were a hospital, an inn, a sanitarium, and other smaller buildings. Why investors would build such a large hotel in a relatively rural and out-of-the-way location is certainly a mystery today. Whatever the reason, the hotel was not a success and it was demolished in 1895. Some sources suggest that a financial panic sweeping England in 1893 cut off funds to the project. The name of the town where the hotel was located is Harrogate. It was named for the city of Harrogate in England and it was founded in 1888 by the same English land company that built the Four Seasons Hotel. Middlesboro, Kentucky, located on the other side of Cumberland Gap in Kentucky, was also founded by this same company in 1890.

The grounds on which the Four Seasons Hotel was built are now occupied by Lincoln Memorial University. The university was founded in 1897, soon after the Four Seasons hotel was demolished. The exact location of the hotel is now occupied by Duke Hall. The university owns 1,300 acres of land in Powells Valley, where it is located, and another 2,000 acres of forested land on Cumberland Mountain, where Gap Cave is located. Gap Cave is still the source of water for the university.

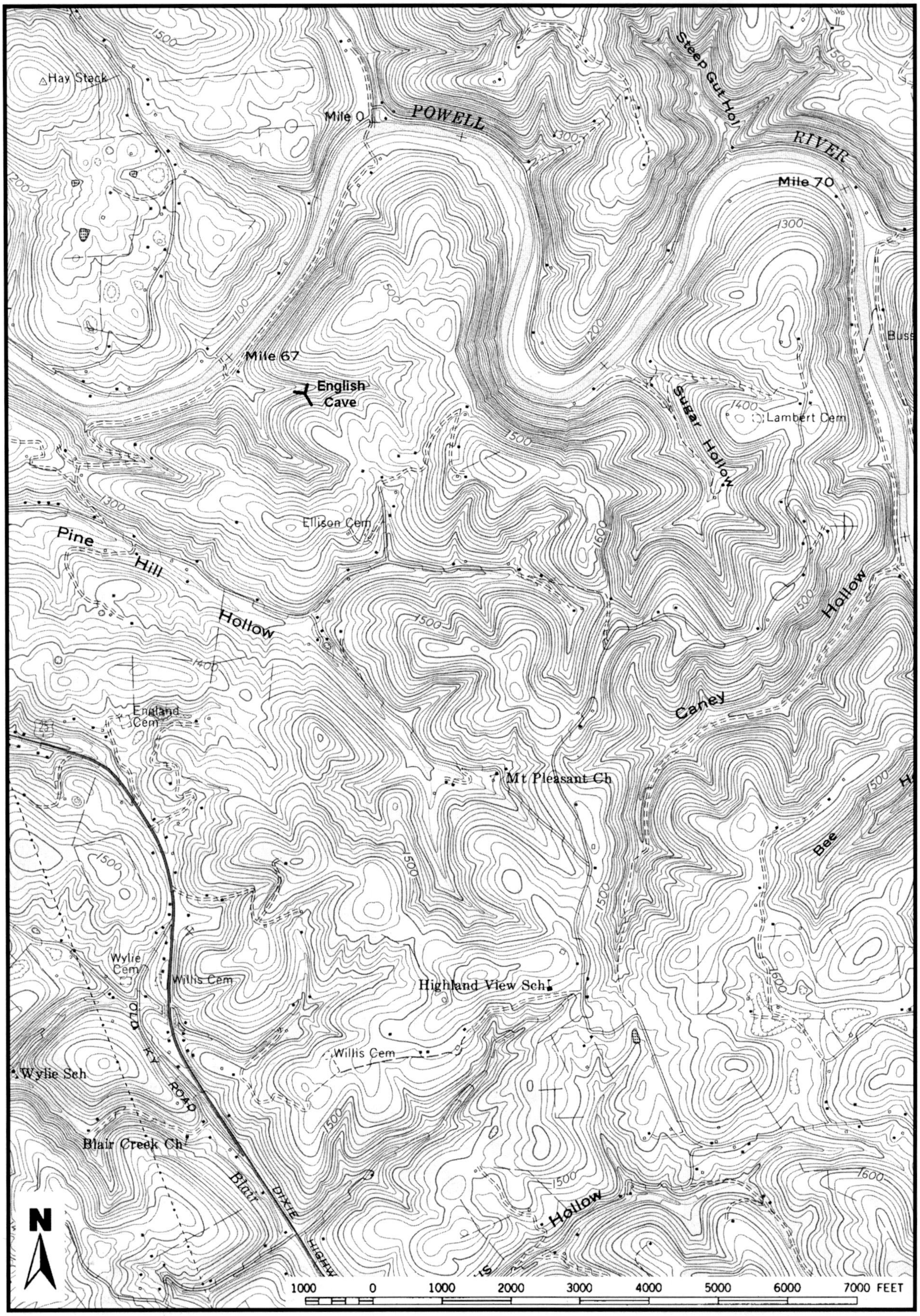

Figure 2.1. Topographic map showing the location of the entrance to English Cave.

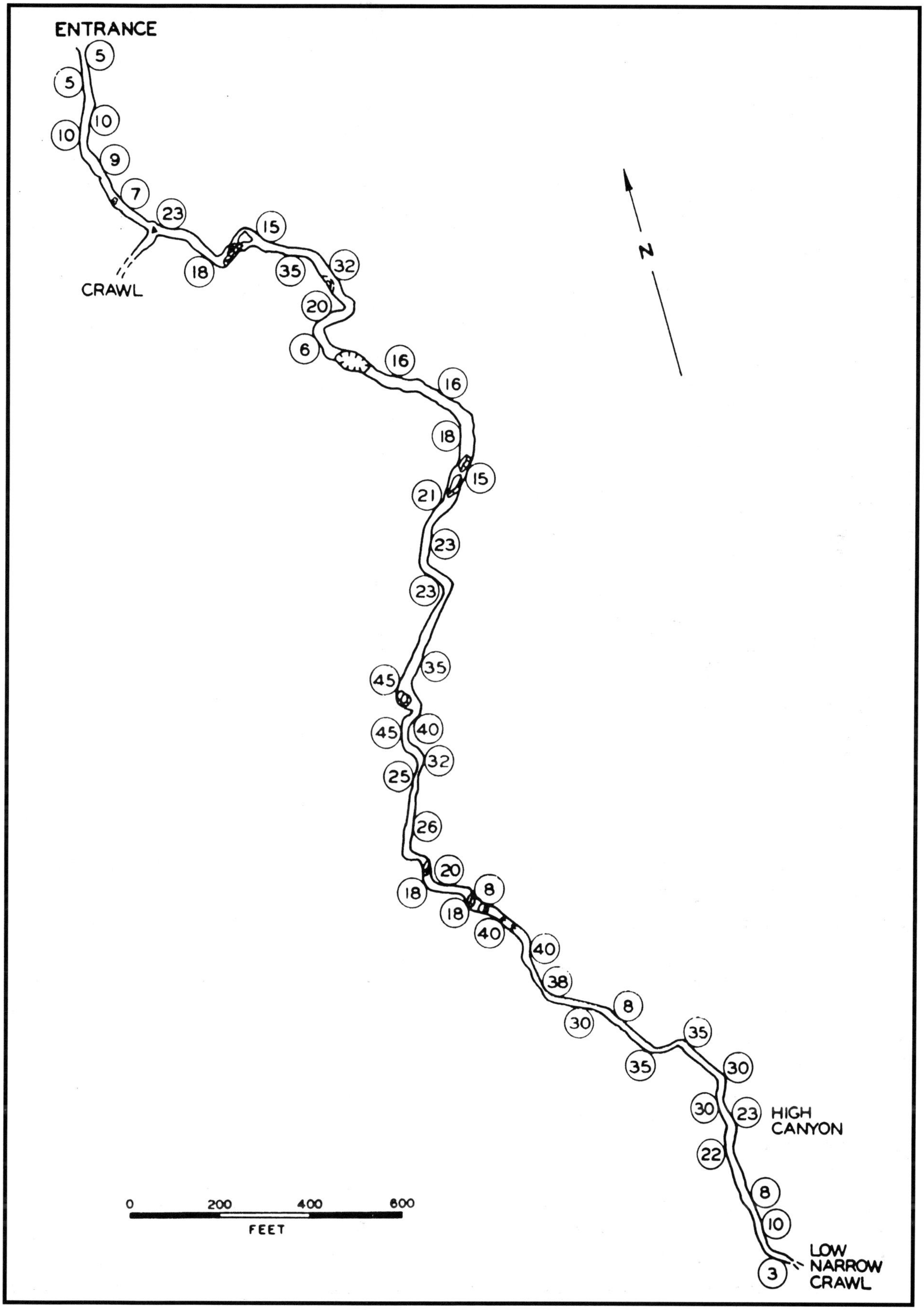

Figure 2.2. Map of English Cave from Caves of Tennessee, *page 116. Mapped by T.C. Barr, B.C. Stewart, and C.E.L. McCary, 1956.*

Figure 2.3. Photograph of Mike Crockett at the entrance to English Cave. Photo by Cheryl Pratt, February 2008.

A Description of the Cave

English Cave is not open to the public at the present time. The book, *Caves of Tennessee*, which was written by Thomas C. Barr Jr in 1961, gives the following description of the cave:

> English Cave is the largest cave visited in Claiborne County. It is developed in gently inclined strata and exhibits strong joint control. The cave trends southeastward through the ridge that forms the interfluve between the cave hollow and Pine Hill Hollow.
>
> The main passage averages 40 feet high and 20 feet wide and is 3,500 feet long. Many formations are developed, including a large, balloon-shaped stalagmite 15 feet high and 8 feet in diameter, a flowstone 20 feet high and 10 feet wide, and a variety of smaller stalagmites. Cave coral incrusts the walls and floor in many places, and two or three extensive breakdowns have formed, over which one must climb to reach the end of the cave. The floor is damp or muddy throughout. Near the end a wet-weather surface stream enters the cave, and a cobble-floored canyon 300 feet long has been developed at this point. It tapers down to a low bedding-plane crawl.
>
> English Cave, formerly owned by the Massengill family, is sometimes spoken of as "Massengill Cave." In the early part of the 20th century it was developed commercially by a utilitarian socialist colony of Englishmen at Cumberland Gap, and it still bears the name "English Cave." The rotting remains of wooden stairways, walkways, and bridges may be seen in many places.[1]

---

1 Thomas C. Barr Jr, *Caves of Tennessee*, Bulletin 64, Tennessee Division of Geology, 1961, pp 115–118.

Another description of English Cave appeared in the *SERA 2004 Guidebook*, which was published by the Appalachian Grotto. That description adds some interesting information about English Cave:

Figure 2.4. Early photograph of a large stalagmite, known as The Haystack from Caves of Tennessee, *page 117.*

> English Cave is very well known in northern Claiborne County as "Massengill Cave." The upside-down triangular-shaped 4-foot by 10-foot stoop opening goes into a popcorn-floored walking passage 10 feet high and 6 feet wide which is actually a side passage for the main borehole, found 500 feet into the cave. At the junction with the entrance passage, you can go right on a flat-floored walking passage for several hundred feet before it gradually lowers to a crawl then ends. But to the left at the junction is 4,000 feet of borehole, a huge lofty canyon averaging 10 to 50 feet high and up to 80 feet wide in places. The terrain ranges from breakdown up-and-down climbs, slippery mud slopes, and flat-floored walking passage. This continues until the ceiling gradually lowers to a very low bedding-plane crawl over cobbles, which is said to eventually connect to Slimey Limey Cave. The cave is very, very photogenic and gapingly impressive. There are several side passages and multi-level loops that can be done. There are lots of formations, many of which have been broken; enormous dripstone columns; and startlingly impressive flowstone curtains. One relatively untouched monolith formation is called "The Haystack" and is 20 feet high and 20 to 30 feet in circumference. It is found in the main trunk not far from where the entrance passage joins. English Cave has been described as the "hardest easy cave you will ever do." The historic artifacts mentioned in *Caves of Tennessee* are now scattered about the muddy floor, partially burned. There is some evidence of the depression-era commercialization attempt.

Figure 2.5. The entrance to Lincoln Memorial University. The Four Seasons Hotel stood on this site in 1892 to 1895. The hotel ran commercial tours of both English Cave and King Solomons Cave (Gap Cave). Photo by Larry E. Matthews, March 17, 2008.

Figure 2.6. Photograph of Brandon Hall by a large stalagmite, known as "The Haystack." Photo by Cheryl Pratt, February 2008.

Figure 2.7. Photograph of Cheryl Pratt in the main passage in English Cave next to wooden boards that are the remains of the old walkway through the cave. Photo by Mike Crockett, February 2008.

Figure 2.8. Photograph of Mike Crockett next to wooden boards that are the remains of the old walkway through the cave. Photo by Cheryl Pratt, February 2008.

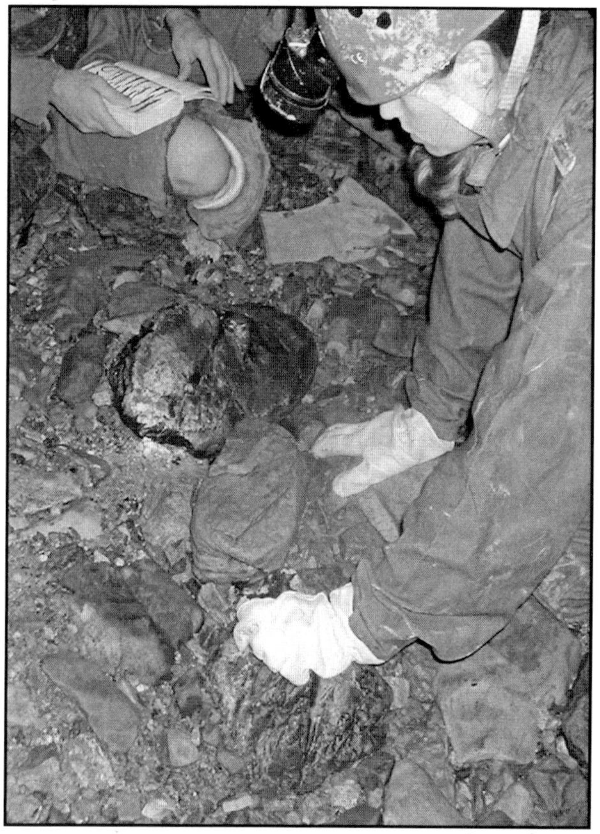

Figure 2.9. Biologist Heather Garland identifying a cave salamander in English Cave. Photo by Cheryl Pratt, 2001.

Figure 2.10. Mike Crockett in the main passage in English Cave. Photo by Cheryl Pratt, February 2008.

Figure 2.11. Brandon Hall next to beautiful cave formations with the cave entrance in the background. Photo by Cheryl Pratt, February 2008.

Figure 2.12. Close-up view of old wooden boards used in walkways. Photo by Cheryl Pratt, February 2008.

There is absolutely no evidence of any saltpeter mining whatsoever as the entire cave is far too damp. This cave is very impressive and is a must-see for first time cavers due to the abundant formations.[2]

The current, mapped length of English Cave is 5,253 feet.

Spelean historian Jim Whidby visited English Cave in 1981, 1982, and 1984. He reports that there are remains of wooden walkways throughout much of the cave. However, much of this wood has now decayed in the damp conditions of the cave. He noted numerous old-style square nails in the trails where the wood had rotted away.[3]

Figure 2.13. Brandon Hall next to a side view of "The Haystack." Photo by Cheryl Pratt, February 2008.

English Cave Today

Currently, English Cave is located on private property and is not open to the public. If you wish to visit this cave, be sure that you have the necessary permission from the owner.

Directions

From Knoxville, go northwest on I-75 for 34 miles to the Caryville–Jacksboro Exit (Exit 134). Turn right on U.S. Highway 25W and proceed 8 miles to LaFollette, Tennessee. Turn right onto Highway 63 and proceed 32 miles to the intersection of U.S. Highway 25E. Turn right on U.S. Highway 25E and proceed southeast for 2.4 miles and cross the Powell River. Immediately turn left onto Pine Hill Road. The entrance to English Cave is approximately 1 mile to the northeast in a small hollow off the Powell River. Since the cave is not open to the public at this time, more specific directions will not be given.

2 *SERA 2004 Guidebook*, the Appalachian Grotto of the NSS, pp 29–30.

3 Jim Whidby, Personal communication, January 25, 2008.

Figure 2.14. The entrance to Cumberland Gap National Historical Park. Photo by Larry E. Matthews, March 18, 2008.

Please Note: This is private property. Do NOT enter this property or English Cave without permission from the owner.

Side Trips

The intersection of Tennessee Highway 63 and U.S. Highway 25E is approximately 2 miles south of Cumberland Gap National Historical Park. With 22,000 acres of spectacular mountain scenery, this park is a must-see. Gap Cave located inside the park is operated as a commercial cave and tours are available on a seasonally varying basis. See Chapter 3 for more information on Gap Cave and Cumberland Gap National Historical Park.

Chapter 3

Gap Cave

The Discovery of Gap Cave

Cumberland Gap is the most important "gap," or passage in the Appalachian Mountains leading west from Virginia into Kentucky and Tennessee. In 1743, Dr Thomas Walker led an expedition for the Loyal Land Company of Virginia as far west as present-day Kingsport, Tennessee. In March 1750 he led another expedition through present day Kentucky that lasted four months. It was during this 1750 expedition that Walker located Cumberland Gap and recorded its existence in his diary. The diary entry for April 13, 1750, reads as follows:

> We went 4 miles to [a] large creek, which we called Cedar Creek, being a branch of Bear Grass, and from thence 6 miles to Cave Gap, the land being levil [*sic*]. On the north side of the gap is a large spring, which falls very fast, and just above the spring is a small entrance to a large cave, which the spring runs through, and there is a constant stream of cool air issuing out. The spring is sufficient to turn [a] Mill. Just at the foot of the hill is a laurel thicket, and the spring water runs through it. On the south side is a plain Indian road .... This gap may be seen at a considerable distance, and there is no other that I know of, except one about 2 miles to the north of it, which does not appear to be so low as the other.

Fur trappers and long hunters knew of Cumberland Gap prior to Dr Thomas Walker, but he is clearly the first person to record this important location. Also, the reference to "a plain Indian road" would indicate an established trail, used by American Indians, hunters, and animals. He also notes the presence of the cave. Since he refers to it as "a large cave," we must assume that he explored it for some distance.

Interestingly, Dr Walker was the physician to Thomas Jefferson's father, Peter Jefferson. After the death of Peter Jefferson, Dr Walker became the guardian of Thomas Jefferson. Dr Walker passed on his love of science and exploration to Thomas Jefferson.

The exploration of the land to the west of the Appalachian Mountains was fueled by more than mere scientific curiosity. According to Katie Algeo:

> Initially, the Euro-Americans who passed through the gap were not much interested in the rugged uplands, but in what lay beyond them: the fertile lands of the Bluegrass, Nashville Basin, and beyond. By the mid-18th century, the rising tide of immigration and the need for new land to replace that depleted by extensive tobacco cultivation and poor agricultural practices created a land hunger that pushed the frontier westward. A few people, such as Kentucky's famed long hunters, crossed the Appalachians on their own, but early mass migration was facilitated by land companies. These companies, usually chartered by politically connected members of the Eastern social elite, received grants for vast tracts of western lands from colonial representatives of the British crown. They sent explorers west to identify promising locations for colonies, entered the claims, then organized parties of settlers. It was in just such a role that Thomas Walker of the

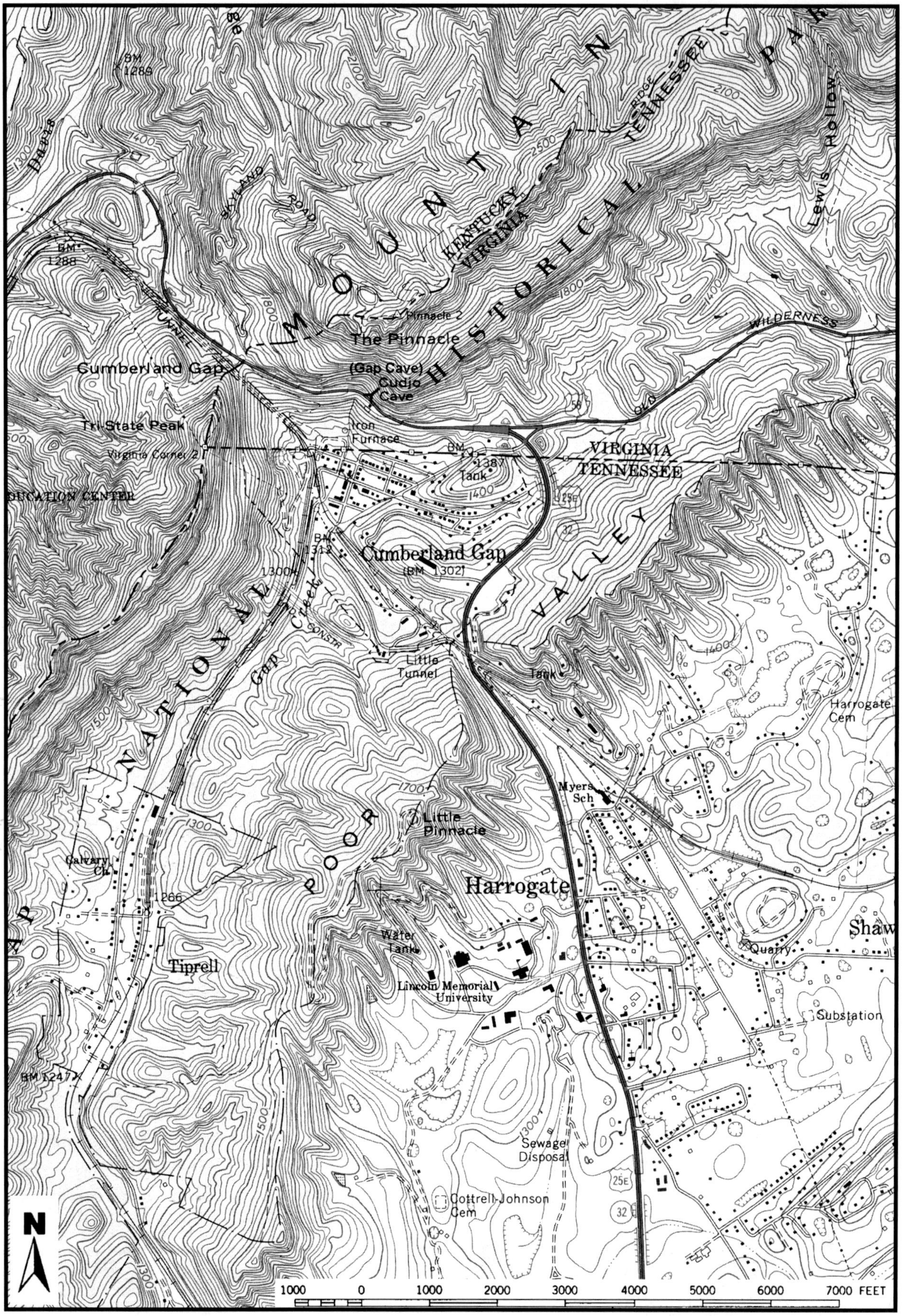

Figure 3.1. Topographic map showing the location of the entrance to Gap Cave.

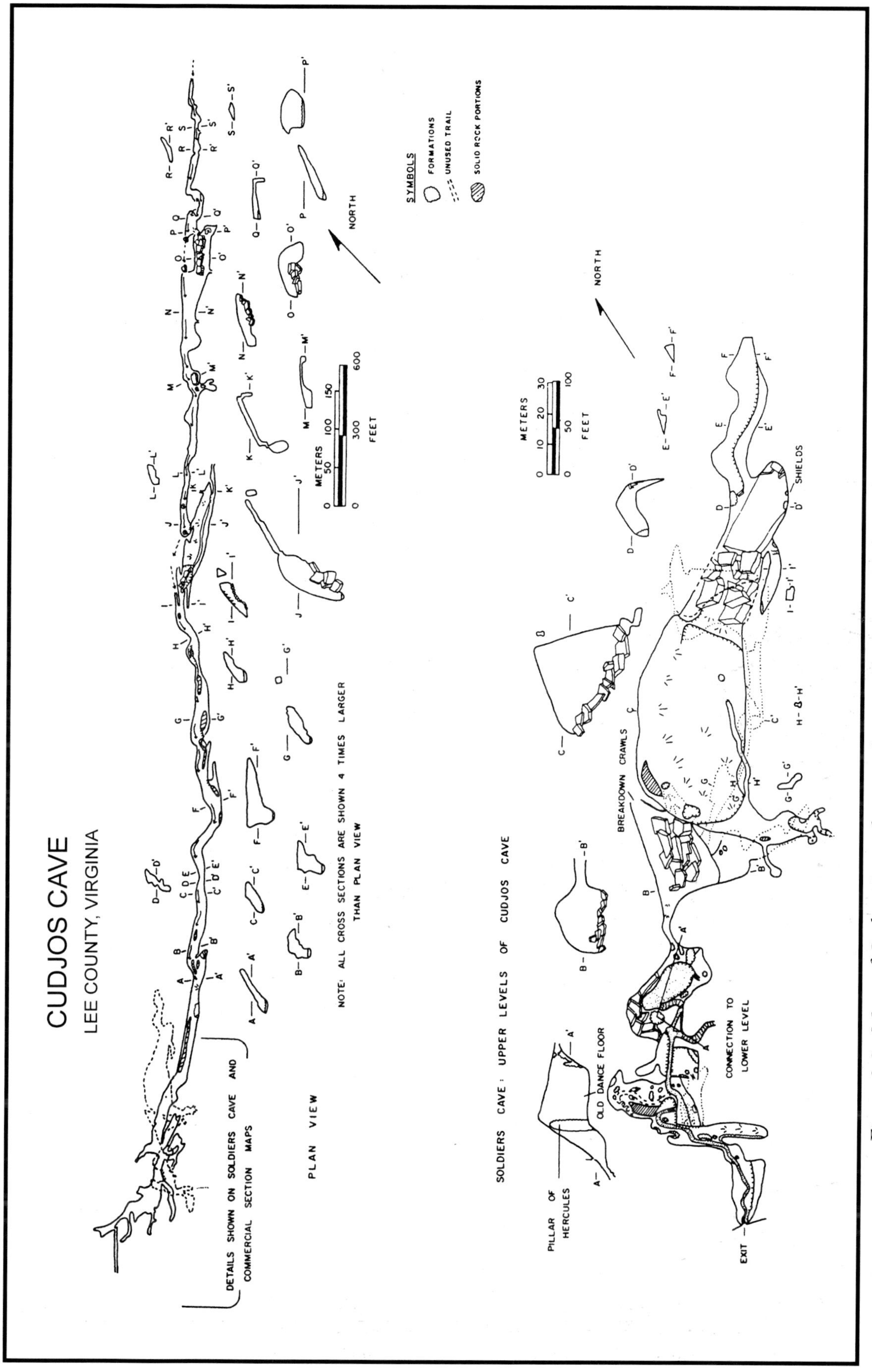

Figure 3.2. Maps of Cudjos Cave, showing the entire cave (top) and the Soldiers Cave section (bottom). From Descriptions of Virginia Caves *by John R. Holsinger (1975). Survey by David Irving and others, 1968–1970.*

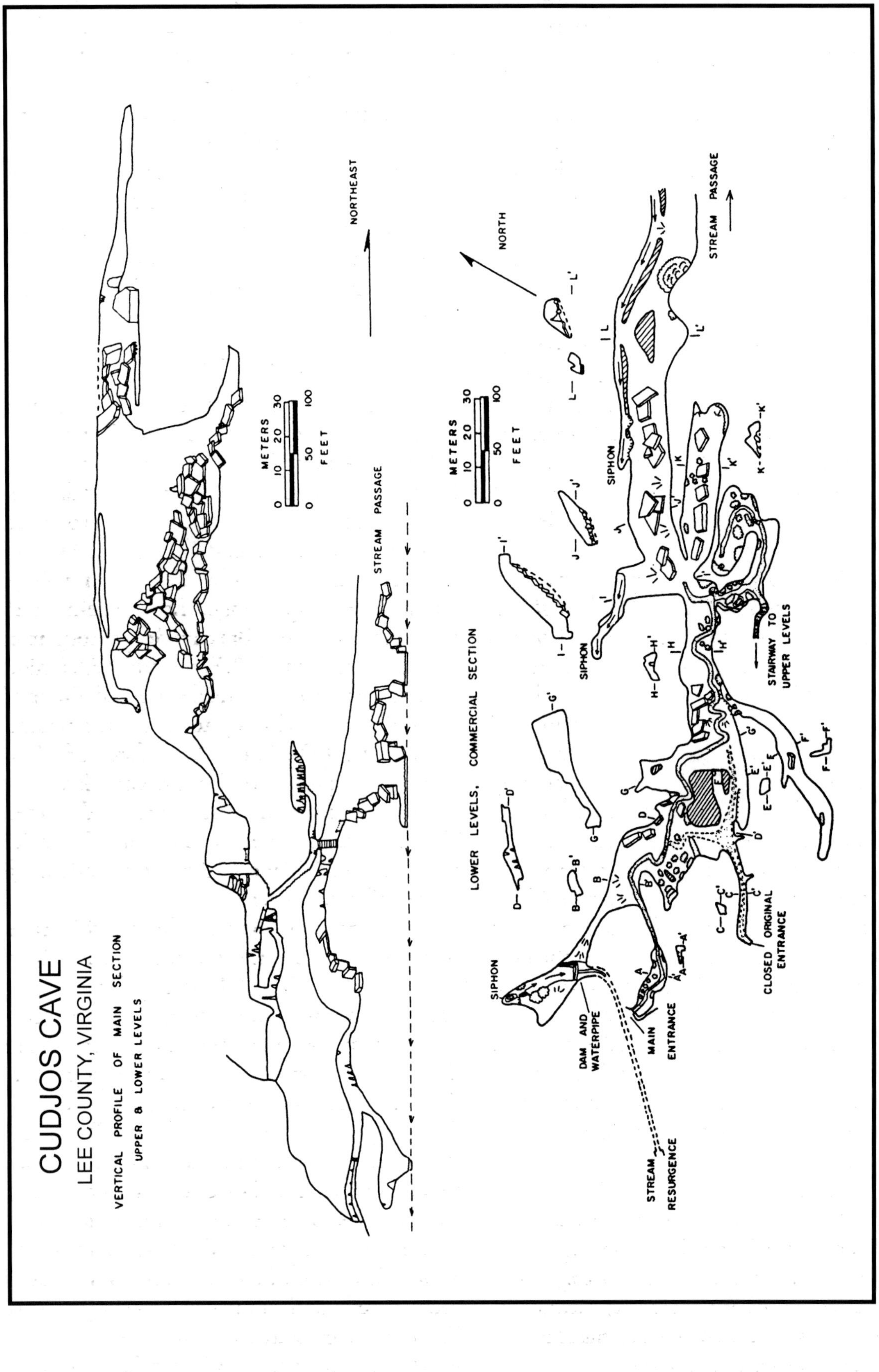

Figure 3.3. Maps of Cudjos Cave, showing a vertical profile of the main section (top) and the lower level of the commercial section (bottom). From Descriptions of Virginia Caves *by John R. Holsinger (1975). Survey by David Irving and others, 1968–1970.*

> Loyal Land Company of Virginia, which had been awarded 800,000 acres, traversed the Cumberland Gap with a small party of explorers in 1750.[1]

Because of its location, the cave was originally referred to as Cumberland Gap Cave. However, Gap Cave is the name currently in use by the National Park Service, who administers the cave.

Daniel Boone and the Wilderness Road

The name most frequently associated with Cumberland Gap is Daniel Boone (1734–1820). Boone was a legend in his own lifetime, as a famous soldier from the American Revolutionary War and even more so as a frontiersman and Indian fighter. He led a party of 30 workmen by following the Wilderness Road through Cumberland Gap and into Kentucky in 1775. A secondary trail branches from the Wilderness Road and became a short-lived artery known as Boone's Path, Boone's Road, or Boone's Trace. The road traversed through the Cumberland Plateau karstlands and cane thickets before opening onto the fertile Bluegrass Region. Exploring into the grassland locality and not far from the palisades of the Kentucky River, Boone established a permanent stockade called Boonesborough. The fort contained one of the first English-speaking settlements west of the Appalachian Mountains. It is estimated that between 200,000 to 300,000 settlers passed through Cumberland Gap on the Wilderness Road and into Kentucky between 1775 and 1810.

Since Thomas Walker noticed Gap Cave during his visit in 1750, it seems reasonable to assume that Boone and the settlers who followed him must have taken notice of the cave entrance too.

---

1 Katie Algeo, "Historical Overview: Settlement History of the Cumberland Gap Region," Published in *Geologic Impacts on the History and Development of Middlesboro, Kentucky*, Kentucky Society of Professional Geologists Annual Field Conference, September 18–20, 2003.

The portion of Gap Cave known as Cumberland Gap Saltpeter Cave shows significant evidence of saltpeter mining. This mining may have occurred as early as the early 1800s, during the time of the War of 1812. This is supported by two signatures with 1812 dates that have been located in the area known as the B Survey. Other industries at Gap Cave used the water power provided by the stream emerging from the lower mouth. This waterpower was used to power a sawmill, a wool carding mill, a granary mill, and the blast machinery for the Cumberland Iron Furnace.

Gap Cave and the Civil War

During the Civil War, Cumberland Gap was considered a strategic location. Confederate Brigadier General Felix K. Zollicoffer occupied it on November 13, 1861. General Zollicoffer and his troops retreated when a larger force led by United States General George W. Morgan moved in and occupied Cumberland Gap. Cumberland Gap changed hands again when Confederate General Braxton Bragg brought an even larger force in 1862. Finally, in 1863, Union General Ambrose Burnside ordered General John Fitzroy DeCourcy to take Cumberland Gap. General DeCourcy led a force of 1,700 men who approached Cumberland Gap from both the north and the south and forced Confederate Brigadier General John W. Frazier to surrender 2,000 men, 18 cannons, and a large supply of ammunition.

Of particular interest during this time period is the fact that the Confederates mined another cave at Cumberland Gap for saltpeter, the principal ingredient of gunpowder. This saltpeter cave was known as Cumberland Gap Saltpetre Cave or, more simply, Saltpeter Cave. Also, at that time, Gap Cave and Soldiers Cave were two separate caves. Therefore, there were three separate caves known at that time: Cumberland Gap Cave, Saltpeter Cave, and Soldiers Cave. A connection was blasted between Gap Cave and Soldiers Cave in the 1930s. The name Soldiers Cave is still used today to refer to the upper entrance to the cave. Then, in 1983, Saltpeter Cave was connected to what was then called Cudjos Cave, to form

Figure 3.4. Lisa Storey by the Soldiers Cave Entrance to Gap Cave. Photo by Ken Storey, December 2003.

soldier stationed at Cumberland Gap described the cave in his diary.

During the Civil War, author John Townsend Trowbridge (1827–1916) published in 1864 a famous book titled *Cudjo's Cave.* This fictional book describes the escape of an east Tennessee slave named Cudjo, who hides in a "Cumberland Grotto" while making his way north to Kentucky on the Underground Railroad. The description of the cave and its location are similar to Gap Cave, so perhaps Trowbridge had visited Cumberland Gap prior to writing his famous book.

Gap Cave Becomes a Commercial Cave

Gap Cave supposedly opened to the public as a commercial cave in 1888. This is not surprising, considering its location was just a few feet off the road that crossed Cumberland Gap. According to one source, the Four Seasons Hotel, located in nearby Harrogate, Tennessee, developed the cave as a tourist attraction in 1892 and renamed it "King Solomon's Cave." For more information on the Four Seasons Hotel, see Chapter 2, page 37. The cave was also known as John A. Murrelll Cave at one time, although the exact dates have not been located.

one large cave system.

Gap Cave and Soldiers Cave were explored by both Union and Confederate troops during the Civil War and some of these soldiers left their names and dates written on the cave's walls and ceiling. There are smoke-drawn pictures of artillery in action and of battle scenes and some of the drawings show flags. According to local historians, the armies used Soldiers Cave as a military hospital and for storage. One Union

Figure 3.5. Civil War era graffiti on a wall in Soldiers Cave showing a cannon. Photo by Cheryl Pratt, May 20, 2004.

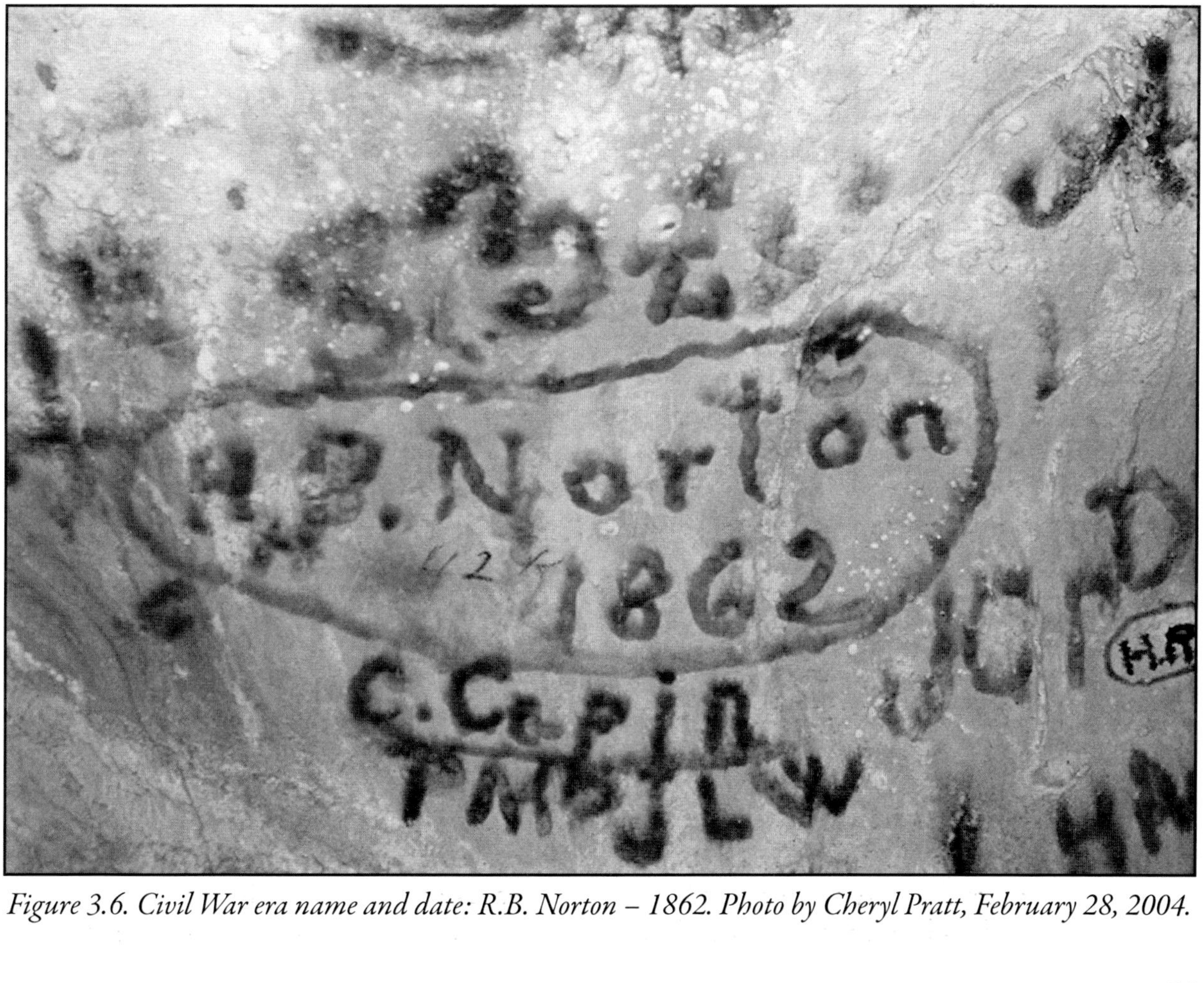

Figure 3.6. Civil War era name and date: R.B. Norton – 1862. Photo by Cheryl Pratt, February 28, 2004.

Figure 3.7. Civil War era name: Francis R. Francher, 11th Mich. Battery. Photo by Ken Storey, November 2003.

John A. Murrell was a legendary bandit, thief, and counterfeiter who was born in middle Tennessee in 1806 and died at the age of 38 in 1844. He operated mostly in west Tennessee and there seems to be no reason to associate his name with Gap Cave.

In 1908, the United States government built a modern gravel road that connected Middlesboro, Kentucky, to Cumberland Gap, Tennessee. This new road went by the Soldiers Cave Entrance to Gap Cave and people were now able to easily drive to the cave in their automobiles. In 1916, Virginia, Kentucky, and Tennessee built additional roads that became part of what was known as the "Dixie Highway," a major tourist corridor through the area. As the road improved, the number of tourists increased.

In 1920, Lincoln Memorial University purchased Cumberland Gap, including Gap Cave and the surrounding land. The college leased the cave to private developers. Due to the popularity of the book *Cudjo's Cave*, the new developers changed the name of the cave to "Cudjo's Cave" to attract visitors. In 1934 electric lights were added to the tour.

The stream in the cave is reported to have supplied both water and power for Lincoln Memorial University at one time and it is still the University's water source today. This water is also bottled and sold under the name of *Cumberland Gap Mountain Spring Water*. The cave was shown under the name "Cudjo's Cave" from the early 1920s until it closed in 1992.

Cumberland Gap National Historical Park

Cumberland Gap was designated a National Historical Park on June 11, 1940. However, at this time Gap Cave was still owned by Lincoln Memorial University. According to the Park Service's Web site:

> The cavern was acquired by Lincoln Memorial University by deed transfer from private owners in March 1920. Title

was conveyed to the commonwealth of Virginia in 1947. The right to operate commercial tours in Gap Cave was transferred to the National Park Service in 1992.

The park consists of 22,000 acres of heavily wooded land, located at the junction of Kentucky, Tennessee, and Virginia. The natural gap in the Appalachian Mountains was used for centuries by migratory animals and American Indians. Beginning in 1775, Cumberland Gap became the primary route for American settlers moving west into Kentucky.

Due to the important history of this area, it has been added to our National Park system. Activities in the park include hiking, camping, and guided tours of Gap Cave.

A Description of the Cave

For most of its commercial history, this cave has been known as Cudjos Cave. However, at the present time the National Park Service is using the name Gap Cave, which they perceive to be the original, historic name for this cave.

Gap Cave is currently open to the public and is shown as a part of Cumberland Gap National Historical Park. Henry H. Douglas gives the following written description of the cave in his 1964 book, *Caves of Virginia*:

> Cudjos Cave is in the very western tip of Virginia, on Route 25E in the west end of Pinnacle Mt, directly above the town of Cumberland Gap.
>
> The cave was named for a colored man who hid in the cave in the early days of the Civil War, where he is supposed to have been killed. It has been variously known since 1750 as "Cumberland Gap Cave," "Gap Cave," "John A. Murrelll Cave," and "King Solomons Cave."
>
> The cave has three entrances and four levels. The commercial entrance leads into the lower level, while the exit is on one of the upper levels. A stream of water flows through the lower reaches of the cave and is part of the water supply of Cumberland Gap. It has many beautiful formations. The length is 1 mile, the depth is 200 feet.

Figure 3.8. Tally marks left by saltpeter miners. Photo by Ken Storey, November 2003.

Figure 3.9. Front of an early commercial brochure for Cudjo's Cave. From the Speleo Archives of Gary K. Soule.

The limestone in which this cave is formed extends in a band all the way to the head of Powell Valley and southwest from there to the vicinity of Sunbright.[2]

2 Douglas, Henry H., *Caves of Virginia*, Virginia Cave Survey, Falls Church, Virginia, 1964, p 318.

Douglas goes on to describe another cave, Soldiers Cave, immediately after the above description of Cudjos Cave, "The dry upper level of Cudjos Cave is generally known as Soldiers Cave. The length is about 480 feet."[3]

Caver David Irving wrote an excellent description of Cudjos Cave in 1970:

> Cudjos is a commercial cave located alongside U.S. 25E just east of Cumberland Gap. The cave is inside the Cumberland Gap National Historical Park, but it is under perpetual lease to Lincoln Memorial University, which operates the commercial cave and uses the stream as a water supply. The present manager of the cave is Mr Seward.
>
> Cudjos was initially two separate caves, King Solomons Cave and Soldiers Cave. During commercialization in the 1930s, a short tunnel was blasted out, connecting the two caves. A new entrance was constructed, and the old entrance to King Solomons Cave was sealed up. The new entrance, right beside U.S. 25E, is the start of the commercial tour. The tour leads eastward through a few formation rooms for 550 feet. There are leads to the left down to the stream and to the right toward the sealed entrance. At the end of this series of rooms is an overlook to the stream. The tour at this point climbs a wooden staircase past a large flowstone formation to an intermediate level. There is always a noticeable temperature increase at the top of this staircase. The intermediate level is very short, but it is extremely well decorated. From this level a blasted tunnel leads steeply up to the main level of Soldiers Cave. Up to the right in Soldiers Cave is the Big Room, which is 100 feet long, 50 feet wide, and 50 feet high. The floor of the Big Room is smooth and it served at one time as a dance hall. At the entrance to the Big Room stands the Pillar of Hercules, a huge totem pole stalagmite which is 10 feet in diameter and 45 feet high. From the Big Room the commercial tour leads southwest

3 Ibid.

Figure 3.10. Mike Stanfill in a high, narrow canyon with boards left by saltpeter miners. Photo by Ken Storey, August 2004.

The Soldiers Cave entrance is high on the hillside, 110 feet above and 360 feet southeast of the main entrance. Another entrance, now gated, is the stream which passes underneath the highway and resurges across the road from the main entrance.

The commercial tour is 1,200 feet long. It is well decorated throughout, although the formations tend to be inactive and have suffered some vandalism. The major part of Cudjos Cave, however, does not lie on the commercial tour. The stream passage may be followed northeast along the strike for over a mile beyond the end of the commercial route. For most of this mile the passage is a large breakdown passage frequently 70 feet wide and 25 feet high. It slants steeply along the dip of the rock. The stream, not very large, lies on the lower left, and a long breakdown-covered slope extends up to the right. There are no extensive side passages along the mile of stream passage. There are several excellent formations in the stream passage, but it is relatively undecorated compared to the commercial areas.

through a formation area (the Cathedral) to the entrance room of Soldiers Cave. This room was used by soldiers quartered in Cumberland Gap during the Civil War, thus giving the upper cave its name. The name Cudjo comes from an old book about a runaway slave who hid in a cave at Cumberland Gap, but the book is fiction, and the cave it describes is completely fictional.

Figure 3.11. Photo from an early commercial brochure for Cudjos Cave showing Three States Corner, Historic Cumberland Gap, Pinnacle Mountain, and an arrow pointing to the cave entrance. From the Speleo Archives of Gary K. Soule.

The other major area not on the commercial route is the upper level of Soldiers Cave. One can scale the far wall in the Big Room and enter a short stoopway which opens into a large breakdown room. The room is 120 feet long, 70 feet wide, and 30 feet high. From here there are a couple of routes through the breakdown into the real "big room" of the cave. This final chamber is a large breakdown room 300 feet long, 100 feet wide, and 100 feet high. There are high leads from each end of the room, accessible from the top of the breakdown. The one to the north, reached by an exposed traverse, is 500 feet long. The high lead on the south end of the room is only 100 feet long, but it intersects an interesting side passage. This is an old stream passage with well scalloped walls which ends abruptly in the middle of the ceiling of the large room. For a very impressive sight, leave a light at the end of this passage and return to the room below to gaze up at the light high overhead. The

Figure 3.12. Postcard showing the ticket office for Cudjos Cave about 1950. From the Larry E. Matthews collection.

Figure 3.13. Old postcard showing the entrance to Cudjos Cave and a car on the highway. From the Gordon and Judy Smith collection at the National Cave Museum, Diamond Caverns, Park City, Kentucky.

Figure 3.14. Postcard showing the "Statue of Liberty" in Cudjos Cave. From the Speleo Archives of Gary K. Soule.

Figure 3.15. Postcard showing "Cleopatra's Bath Tub" in Cudjos Cave. From the Speleo Archives of Gary K. Soule.

Figure 3.16. Postcard showing "Lovers' Leap" in Cudjos Cave. From the Speleo Archives of Gary K. Soule.

upper level of Soldiers Cave is undecorated except for the two high leads from the last big room that have many formations, including two shields.

The total length of Cudjos Cave is just over 2 miles. It has been mapped by crews led by David Irving, but the final map has not yet been drawn. A very short description of Cudjos appears in *Caves of Virginia* by Douglas.[4]

4 David Irving, "A Guide to the Caves of the Oak Ridge – Knoxville Area," *The Speleotype*, vol 5 (1970), nos 1, 2, and 3, pp 3–18, 21–34, 37–52.

Figure 3.17. Postcard showing the "Crystal City" in Cudjos Cave. From the Speleo Archives of Gary K. Soule.

Figure 3.18. Postcard showing the "Capitol Dome" in Cudjos Cave. From the Speleo Archives of Gary K. Soule.

Figure 3.19. Postcard showing the "Pillar of Hercules" in Cudjos Cave. From the Speleo Archives of Gary K. Soule.

David Irving would soon pioneer modern exploration in the caves of Cumberland Gap. Some of those exciting discoveries are described below.

New Discoveries

Members of the National Speleological Society have been mapping and exploring Cumberland Gap Cave for the past 39 years and the current surveyed length of the cave is 13.5 miles (21.8 kilometers). This makes Gap Cave the 53rd longest cave in the United States.[5] The cave system has six entrances: (1) the original entrance near the road, which has been known as the King Solomons Cave Entrance, (2) the artificial Cudjos Cave Entrance, (3) the Stream Entrance, (4) the Soldiers Cave Entrance, (5) the Big Saltpeter Cave Entrance, and (6) the Wellhole, a vertical entrance. Exploration and mapping continue, even as this book goes to press. According to Chuck Wiley and Michael Doughty:

> Gap Cave, formerly known as Cudjos Cave, and the connected Big Saltpeter Cave, have been surveyed by generations of East Tennessee Grotto members starting with Dave Irving and Jeff Bowers in 1969, and peaking with the connection of the two caves in 1983 and the survey trips that followed. Gap Cave is now within the Cumberland Gap National Historical Park, located on the border of Tennessee, Virginia, and Kentucky. Activity in the caves of Cumberland Gap was in a lull during the late 1990s as ownership changed hands. Finally a new generation of cavers stepped up to deal with the "administrivia" of the National Park Service. The Cave Research

5 Bob Gulden, *USA Long Cave List*, January 30, 2008.

Figure 3.20. Postcard showing both the Cudjos Cave Entrance and Souvenir Shop. From the Speleo Archives of Gary K. Soule.

Foundation (CRF) is coordinating the current project at Gap Cave, with Mike Crockett as Project Coordinator. East Tennessee Grotto's involvement began in earnest in 2003, and over the next three years 23 members supported the project during 39 trips.[6]

David Irving wrote a trip report in 1969

6 Chuck Wiley and Michael Doughty, "The History of East Tennessee Grotto," *The Speleotype*, vol 39 (2006), no. 2, pp 29–45.

Figure 3.21. Postcard showing tourists on the trail in Cudjos Cave. From the Speleo Archives of Gary K. Soule.

describing one of the very first trips in the modern exploration of Gap Cave. His report on Cumberland Gap Saltpeter Cave is reprinted below:

Figure 3.22. Huge breakdown room, the Coke Bottle Room, with light from an upper level passage entering near the top. Photo by Ken Storey, May 2004.

> The Saltpeter Cave is the second largest cave in Cumberland Gap National Historical Park, being surpassed only by Cudjos. The name is mentioned in Douglas's *Caves of Virginia.* Unfortunately, the cave described there is not Cumberland Gap Saltpeter Cave, but a small cave known as Skylight Cave. Cumberland Gap Saltpeter Cave is described, complete with a sketch map and several photographs, in a report on caves in Cumberland Gap National Historical Park by Andrew Loveless, a former park historian. This report also contains several interesting comments on the saltpeter mining in Cumberland Gap Saltpeter Cave as well as descriptions of other caves in the park. Unfortunately, it is available only at the park headquarters near Middlesboro.
>
> We first became interested in Cumberland Gap Saltpeter Cave after mapping the stream passage in Cudjos Cave. The map showed that we had turned back underneath a small surface hollow about 1,500 feet west of the entrance to Cumberland Gap Saltpeter Cave. At first we hoped that Cumberland Gap Saltpeter Cave would lead down to the Cudjos stream and possibly into Cudjos Cave at the upstream end. This, however, does not seem possible now. The entrances of Cudjos Cave and Cumberland Gap Saltpeter Cave are 400 feet apart vertically. Following the

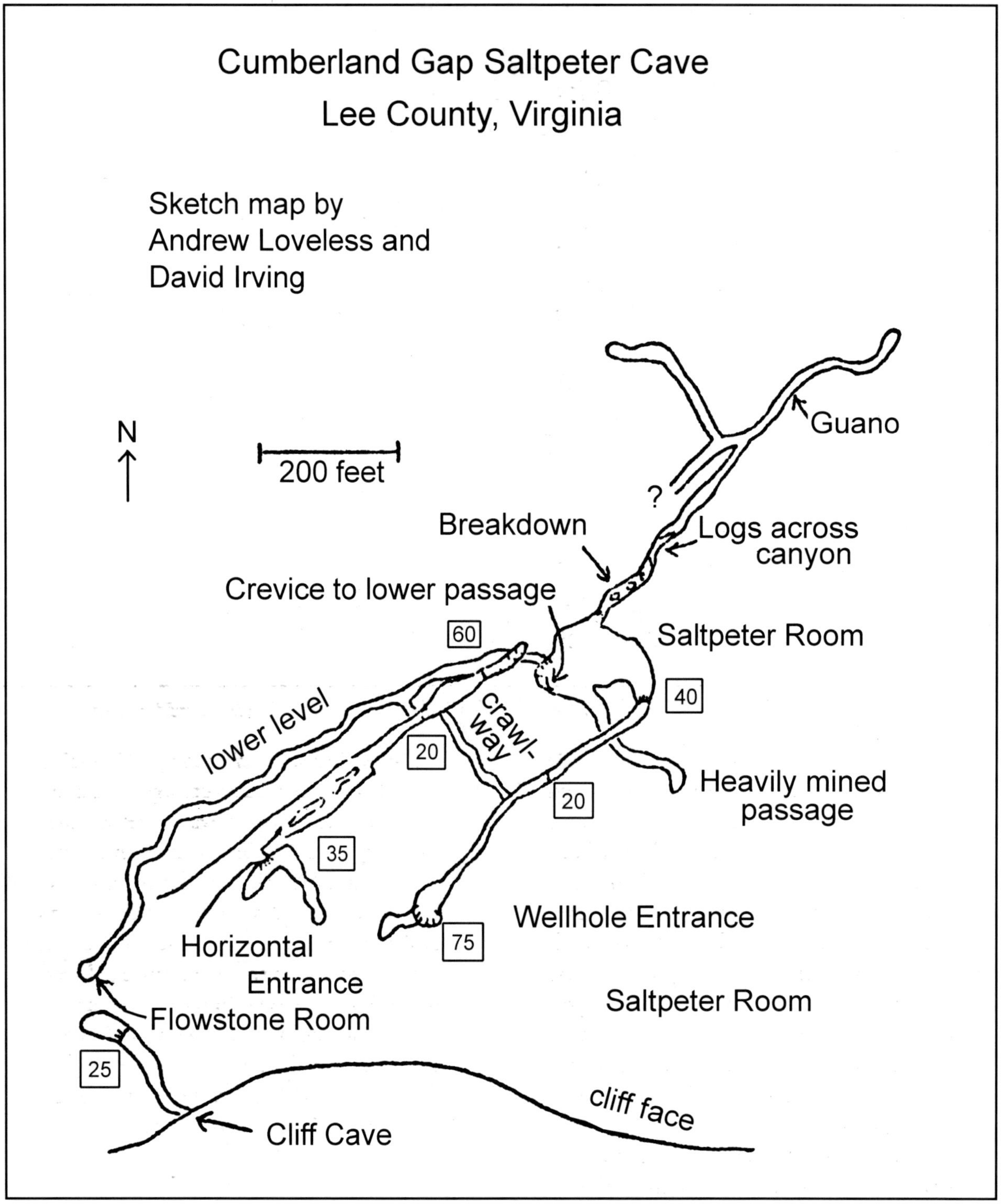

Figure 3.23. Sketch map of Cumberland Gap Saltpeter Cave (1969) by Andrew Loveless and David Irving.

stream in Cudjos, we gain only 50 feet in a mile. We have descended at most 200 feet in Cumberland Gap Saltpeter Cave with little prospect of going substantially deeper. In addition, Cumberland Gap Saltpeter Cave heads northeast along the strike and away from Cudjos Cave.

Cumberland Gap Saltpeter is an interesting cave on its own and well worth a visit. Our first trip in late December consisted of Francis McKinney, Pete Hauer, Jack Speece, Jim Tunnell, Chuck Wiley, Mike Skorski, Roger Johnson, and me. The weather was cold, wet, and thoroughly miserable—the sort of day when you wonder why the hell you ever

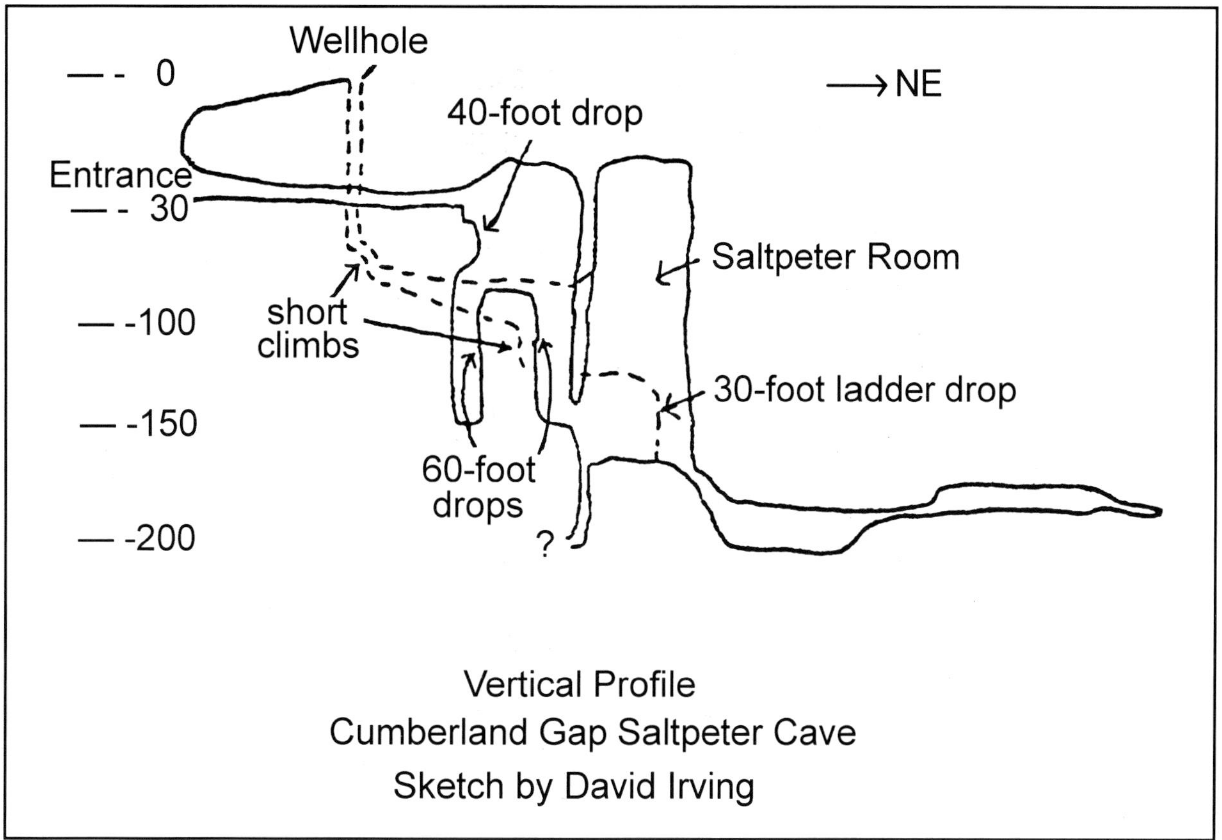

Figure 3.24. Vertical profile of Cumberland Gap Saltpeter Cave (1969) by David Irving.

took up caving anyway. We parked at the Pinnacle, overlooking the Gap, and hiked along the Ridge Trail to a point above the cave where we cut downhill. Without much trouble we located the Wellhole, or vertical entrance. We rigged a rope and rappelled down the 75-foot entrance shaft. McKinney found and is keeping the first carbide lamp he has ever owned. A canyon passage with several short climbs leads to the northeast. At a 30-foot drop we rigged a cable ladder. The bottom of this drop opened into the Saltpeter Room, a large room containing the remains of the saltpeter mining operations in the cave. Pete Hauer promptly went wild examining the remains of vats and searching for the signs of digging in a nearby passage. The Saltpeter Room, the intersection of several passages at different levels, has high dome-like walls rising 125 feet to the ceiling. We climbed down through breakdown and explored the lower levels. Upstream I quickly found myself straddling over rimstone pools while ducking under bat colonies (*Myotis gricescens*?). Deducing that this was a winter colony, I decided that this passage could wait until summer to be explored. In the downstream canyon we found more relics in the form of log bridges across drops and Bill Cuddington's NSS number. He must have been lost to get that far from the vertical entrance. We had to leave the cave early in order to return to the Pinnacle before the gates were closed at sunset. As we climbed up the steep hill to the trail, in light snow and sleet, we vowed to find an easier route to the cave.

* * * * * * * * * * *

The second trip was made March 30 with Francis McKinney, Roger Johnson, John Clarke, and me. This time we parked near the Lewis Hollow picnic area and hiked the trail up Lewis Hollow. When we were above the limestone, we bushwhacked westward atop the limestone bluffs to the Saltpeter Cave. We easily found the horizontal entrance and marveled at how we managed to miss it the first time. This

Figure 3.25. Front and back of a commercial brochure for Cudjo Caverns – Cumberland Gap National Park. From the Speleo Archives of Gary K. Soule.

entrance is triangular shaped, 25 feet high, and 20 feet wide at the base. It is located in a large sink just 200 feet west of the Wellhole.

We soon discovered that the horizontal entrance is worse than the vertical one. The passage quickly narrowed and, less than 300 feet from the entrance, dropped off 40 feet through a narrow fissure. We rappelled down but also rigged a cable ladder for the return because the fissure was so tight at the top that prusiking would have been miserable. At the bottom of this drop, we found ourselves on a short platform with 60-foot drops at either end. Loveless' report said to descend the further drop, but we

Figure 3.26. Photograph showing the upper commercial entrance to Gap Cave and the lower Stream Entrance. Photo by Ken Storey, June 2003.

couldn't find a decent rig point; so I tried the near drop. It was dead-end except for a low crawl. Roger came down and crawled through, but I couldn't make it without digging. Frustration! I began to prussic up, but before I reached the top, Roger had found a connection to the other 60-foot drop, located a climbable chimney, and was waiting at the top for me. Frustration squared.

We finally found a rig point—through a hole in the ceiling—for the 60-foot drop, but decided that Roger's chimney was the best route. Getting started into the chimney takes guts or stupidity or both. You're sitting on a sloping ledge and have to leap into the chimney, praying that you can wedge yourself in before you hit the bottom. The bottom of the 60-footer leads quickly to a 15-foot climb down into the Saltpeter Room. We again explored the lower level passages leading northeast out of the Saltpeter Room before leaving the cave.[7]

Perhaps one of the most exciting discoveries since this new era of exploration began was the connection of Cudjos Cave to Cumberland Gap Saltpeter Cave on July 30, 1983. Below, is a description of that trip as recorded by Jeff Bowers:

---

7 David Irving, "Cumberland Gap Saltpeter Cave, Lee County, Virginia," *The Speleotype*, vol 4 (1969), no. 4, pp 67–71.

On Saturday, July 30, 1983, at or about 8:00 P.M. the long sought after connection of Cumberland Gap Saltpeter Cave to Cudjos Caverns was finally realized. The two caves have been hydrologically connected for years, but no one had made the physical connection, even though many attempts by many cavers had been made over the years. Dave Irving, former East Tennessee Grotto Chairman, had begun a project on the two caves in the late 1960s.

Figure 3.27. Photograph showing old stone steps and massive formations above the Ballroom in the commercial section. Photo by Lisa Storey, October 2003.

Those along on the July 30 trip were: Marion Smith, Gerald Moni, Ed Halladay, Brian Ball, Mike Green, Jim Richards, Sarah Gayle, and the author (Jeff Bowers). All but Jim and Sarah made the actual connection; they had turned around only one hour before the breakthrough.

At this point some background on recent events which led up to the connection is necessary. My first exposure to the Saltpeter Cave was in July 1982, during the SERA Cave Carnival. The cave was a featured trip and Richard Wallace, cave biologist extraordinaire, led a large party to the cave. Both the horizontal and Wellhole entrance (75-foot pit) are high on the southeast slopes of Cumberland Mountain, above and northeast of the Pinnacle, the Wellhole at an elevation of 1,970 feet. It is the more convenient access to the cave, as the horizontal entrance is fenced 200 feet inside to prevent entry during the bat season, as both Indiana and grey bats use the cave. Also, the horizontal entrance leads to a 40-foot drop, a rather exposed 60-foot drop, and then a 12-foot climbdown into the Saltpeter Room, a large dome complex 100 feet long and over 100 feet high, with rotted remains of vats. The

Wellhole route involves the initial drop, a tight squeeze, and a sloping canyon about 300 feet long with several short climbs, [a] 10-foot chimney, and a 35-foot rope or ladder drop down the canyon into the level of the Saltpeter Room. This drop is rigged by tying off to a boulder, crawling under it, then traversing out the canyon about 10 feet and slipping into a bolt to permit a clean drop between the two walls.

Phreatic canyons lead out of the room in several directions, but the major one leads northeast into a multi-level canyon with stout timbers wedged in the passage at several points. The cave is dry at this point, and these timbers are so well preserved they are safe to walk on. After about 800 feet, a short crawl leads to a junction of several dry canyons. The correct lead goes about 150 feet until a low, wide, sandy crawl is reached. (The passage also continues straight ahead.) After about 150 feet of belly crawl in this breezy passage, it opens up into a multi-level canyon lined with occasional gypsum and dried up speleothems. This is the B survey. After a couple [of] thousand feet of walking, stooping, and crawling, a rather formidable breakdown barrier is reached at station B-100. One is now headed southwest, as the passage makes a long 180-degree turn. This is not at all apparent without a compass.

* * * * * * * * * * *

On a trip the second weekend in September 1982, Marion Smith, Art Cathers, Richard Wallace, Chuck Wiley, Joe Parrot, and I pushed to the left in this breakdown, down through fissures and more breakdown until, after losing perhaps 75 feet elevation, we reached the large void of the stream passage (40 feet high and 80 feet wide). After traveling a short distance downstream, a huge breakdown stretching to the ceiling forced us to climb back down to the stream. But after a short distance we were stopped. A way was found up through this mess into a large breakdown room/passage: the H survey. At H-19 (100-foot shots), the passage terminates. This is the last surveyed point in the cave.

Backtracking, Joe and I found a lead going down again. Marion, Art, and Mike disappeared into this and didn't show up at the entrance until 12 hours after everyone else. They had forced their way down through breakdown to the stream again, but only penetrated perhaps 200 feet before reaching another formidable barrier. They exited after 14 hours in the cave.

* * * * * * * * * * *

On Saturday, July 16, a return trip was made, this time with Dave Irving

Figure 3.28. Black salamander with small, white spots in Gap Cave. Photo by Cheryl Pratt, August 28, 2004.

Figure 3.29. Snake inside Gap Cave. Photo by Cheryl Pratt, July 26, 2003.

Figure 3.30. Massive flowstone in Gap Cave. Photo by Cheryl Pratt, March 28, 2007.

After traversing the breakdown toward the center of the room, we intersected tracks and point H-2. We had simply come into the top of the room near the point where the Wilderness Road takes off, this being the only passage going under Kentucky. Upon returning to B-100 after wasting two hours, Dave casually mentioned it was only seven stations through the breakdown from B-100 to F-7. We followed it and straddled a 40-foot pit. At this point Gary was climbing up a crack between two large boulders. I perceived that one was moving and decided to make myself scarce. At F-6 you pop out of the breakdown right beside another

himself (still around from the NSS National Convention), Richard Wallace, Art Cathers, Gary Daughtery, and me. Upon arriving at B-100, it was decided to probe the breakdown in other directions to see what we could find. Richard and I probed off to the right and after twisting around and up through several rooms in the breakdown, popped through a squeeze into an enormous room. We could not see the bottom or the other side. We were quite baffled at this point, as the area appeared virgin and we could see no tracks.

large pit with jagged rocks at the bottom. This slope is rather exposed and requires much caution. Gary and Richard somehow were at the bottom of the room—two tiny figures. The enormity of the breakdown hit home: you can lose or gain over 100 feet in elevation just climbing breakdown!

Our goal was H-19 and a lead Mark Stock found in 1978. He had solo-explored perhaps 500 feet of virgin passage while the others waited. The lead is not apparent until you are within 2 feet because the approach is belly-crawling over large slabs

Figure 3.31. The Coke Bottle Room. Photo by Lisa Storey, January 2004.

Figure 3.32. Flowstone-covered Coca Cola bottle in the Coke Bottle Room. Photo by Lisa Storey, September 2003.

100 feet wide and 10 feet tall, with several massive pits in the floor. Dave told us of a route to an upstream section where over a kilometer of stream had been seen but not surveyed. He also reminded us this was only the fourth trip into the H survey.

After reaching Mark's lead, Richard and Gary slid down a sloping crevice a foot or two wide with no apparent bottom. After losing perhaps 75 feet of elevation, they reached a rather exposed climb. It was only 8 feet but the exposure was 20 feet. Art and I rigged a handline to a flake and everybody was in Mark Stock's passage. H-19 plots up within 50 meters of upstream Cudjos, so we knew we were close. We explored old broken-down stream passages and probed many holes in breakdown. The air was elusive and hard to follow. Finally we reached the limit of Mark's exploration and penetrated into the unknown. My Suunto said southwest. Go for it! We had

Figure 3.33. Richard Knapp by formations and a rimstone pool. Photo by Cheryl Pratt, August 28, 2004.

lost Art and Gary but weren't about to stop, as we had found our way into a dip slope trunk 60 feet wide with flowstone and rimstone. After a few hundred feet it terminated. At this point I was desperate, probing under ugly, muddy breakdown. We had lost the air. Eleven hours into the cave, we decided to head out. Art and I reached the surface after 16½ hours underground, but had to wait for the others two hours at the entrance.

* * * * * * * * * *

Marion caught wind of this and decided a quick follow-up was in order, so we once again applied for a permit. By now I was familiar with the 1½ mile hike up Lewis Hollow Trail and the escarpment, a gain of 800 feet elevation. I led the troops to the Mark Stock lead, as I was the only one who had been there. Using every possible shortcut, we were at the bottom of the crevice after only 4 hours. We reached our previous explorations and exhausted every possible lead before backtracking. At this point my electric and carbide were both acting up (twink, twink, pffft!). Back in the walkway stream passage below the crevice, Marion crawled over some breakdown and found a large amount of air going down it. After a brief recon, he came back for us and we pushed on.

After only 200 feet of stooping and crawling through virgin cave, Ed climbed down 6 or 8 feet and hollered, "Footprints!" We went a little farther and there was the stream. A short distance upstream was a sump. Then an "out" arrow, pointing downstream. There was hollering and screaming, but I was overcome with humbleness for all those who had tried and failed.

Now, we'll be out soon. Right? Wrong! None of us had been in Cudjos before and didn't realize how far it was to get out.

Figure 3.34. Richard Knapp in rappel with a duffel bag. Photo by Cheryl Pratt, May 15, 2004.

Clambering over huge slabs, wading in the stream, going up and down to find a good route, the passage became more impressive the farther on we went. Passing some large rimstone dams and flowstone mounds, the passage became perhaps 80 feet wide and 50 feet high.

At several points we climbed up to dry leads with many signatures from the 1920s and 1930s but still none as old as the two 1812 dates in the B survey in the upper cave. They had penetrated very deep for their day. Finally, after well over a mile, we saw the telltale electric wires and a steep trail up the breakdown to the end of the tourist trail. We toured the very impressive commercial section with massive formations as much as 35 feet tall (Pillar of Hercules). We ascended the wooden steps to the level of Soldiers Cave.

Cudjos was originally two caves: King Solomons and Soldiers. A connection was blasted between the two in the 1930s; the King Solomons entrance was sealed; and a new entrance was opened right beside U.S. 25E. At the Soldiers Cave entrance we went out and decided it was too hot. We had been in the cave only 11 hours at this point.

Marion, Ed, and Brian decided to go back through the cave and I managed to talk Ed into picking up my vertical pack at the end of Cudjos. He had no choice since his car keys were in his pack in the same place. Gerald, Mike, and I toured some more cave, exiting at the Cudjos entrance right under their noses. Now was not the time to tell them all about it. We walked the 2 or so miles back to the campground. I had told Marion and Ed I would walk back up the mountain and meet them in the Saltpeter Room, as I had stashed gear and food there and at the entrance. They hadn't believed me, so

Figure 3.35. Cheryl Pratt by beautiful flowstone formation in Gap Cave. Photo by Ken Storey, April 2005.

their time getting out. They wondered how I had passed them and I told them the news. At the entrance I put on the seat and rack I had borrowed from Mike and descended to the Saltpeter Room, where I waited one whole carbide burn beforebecomingtoo cold. I improvised a Texas system with the slings on hand and went back to the base of the Wellhole, where I fell into a deep sleep. Finally, they showed up. They had become confused in the H survey. The time was 4:45 A.M. They had taken 6½ hours to come back through. I left them a note in the Saltpeter Room so they knew I wasn't lying. The trip had taken its toll in lost and trashed gear: my 3-year-old Bluewater II had been severed to the core where it lay on the bottom of the Wellhole, creating a 50 and a 150. But it was well worth it. Ed, Brian, Mike, and I reached Knoxville at 8:30 that morning.

I wasn't about to let them make me out a liar. Besides, I was responsible for my gear and didn't want to leave it.

At the trail turnoff halfway up I met Jim and Sarah, who apparently had taken

* * * * * * * * * *

The work is far from over. The

Figure 3.36. Lisa Storey by the gate at the commercial entrance to Gap Cave. Photo by Ken Storey, July 2003.

> connection makes the combined depth of the cave approximately 450 feet and combined length of over 4 miles. There is much unsurveyed passage, at least in the Saltpeter Cave. There are even unsurveyed leads in the B-survey. Now that the connection has been made, Dave Irving will probably pass the project on, something that must be hard for him, as the project has been his for over 10 years. It must be a heartbreaker to know we were within 200 or so feet on the previous trip. Even Mark Stock was as close as anybody had been.[8]

Not only is Jeff Bowers' account of the discovery trip exciting to read, but also it conveys to the reader some understanding of the complexity and size of this truly fascinating cave.

8 Jeff Bowers, "Connection At Last!," *D.C. Speleograph*, vol 39 (1983), no. 9, pp 3–5.

The Cumberland Gap Tunnel Caves

Prior to excavating the major highway tunnel under Cumberland Gap, a 10-foot by 10-foot pilot tunnel was drilled to determine if there would be any unforeseen problems that would affect the construction of the tunnel. Chuck Wiley and Michael Doughty describe these efforts to restore old Route 25E to its conditions as would have been seen during pioneer times:

> About this time (1987), the East Tennessee Grotto got involved with the Cumberland Gap Tunnel Caves project for the National Park Service. North- and south-bound tunnels were being bored through the mountain at Cumberland Gap. The highway would then be shifted to the tunnels, and the old route over the gap, passing by Cudjos Cave, would be restored to the original foot trail used by Indians and pioneers. The 10-foot by 10-foot pilot tunnel intersected two caves, and the Park Service and tunnel contractor

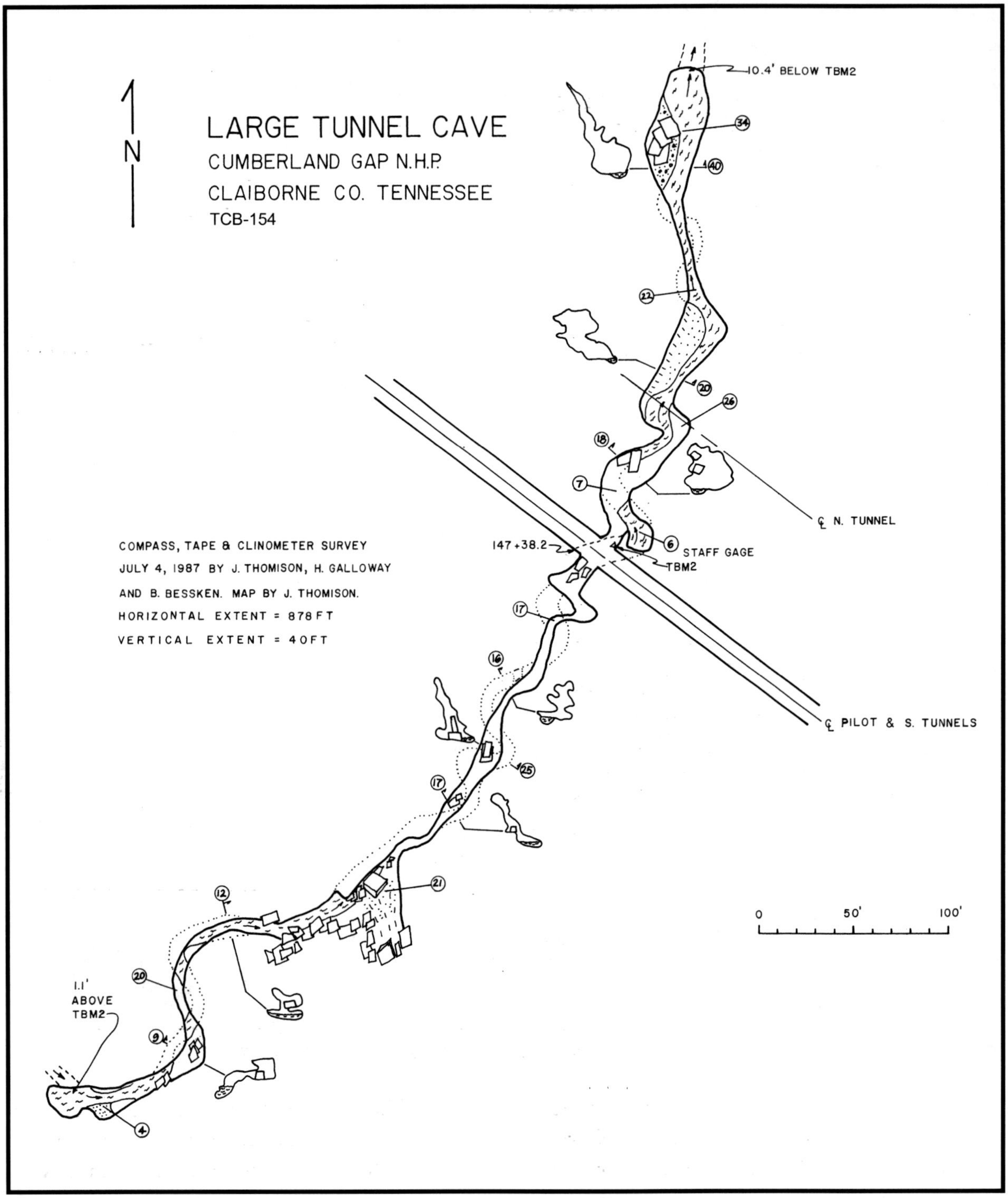

Figure 3.37. Map of Large Tunnel Cave. Map courtesy of Jack Thomison.

were interested in knowing the extent of these caves. East Tennessee Grotto members explored and surveyed the caves and provided maps to the National Park Service. Mike Doughty and Jack Thomison did a bolt climb in an effort to follow any upper leads. There was much ridge walking done on the surface over the caves.[9]

Caver Jack Thomison provides the following descriptions and maps of these two caves:

The first cave, located about 1,272 feet

9 Ibid.

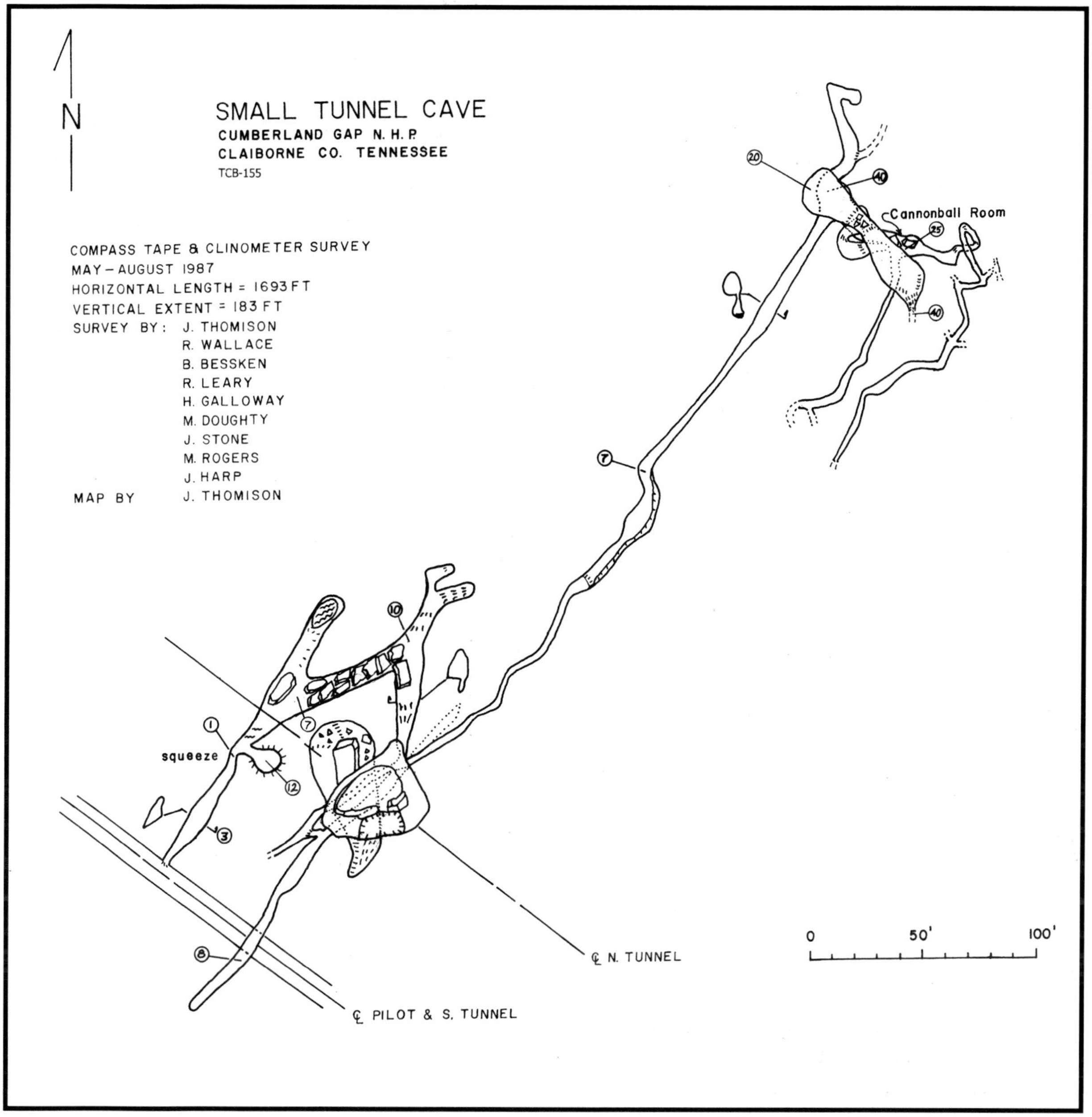

Figure 3.38. Map of Small Tunnel Cave. Map courtesy of Jack Thomison.

north of the Tennessee portal, appeared the larger and was called Large Tunnel Cave. The pilot bore intersected a walking-size stream passage in Large Tunnel Cave. The other cave was a blowing crawl, considered insignificant by the Federal Highway Administration. It was located about 1,010 feet north of the Tennessee portal and was called Small Tunnel Cave.

The National Park at Cumberland Gap National Historical Park and the Federal Highway Administration contacted cavers of the East Tennessee Grotto to explore and map the caves in May 1987.

The blowing lead in the pilot bore for the Small Tunnel Cave continues east through a tight dug crawl to a muddy room at about the proposed location of the northbound tunnel. A passage goes left about 40 feet to a pool and another goes right up a breakdown slope to end in mud fill. A passage goes right before the top and continues up to an intersection. The intersection is about 50 feet above the bottom of the pilot bore and directly above the proposed northbound tunnel. From there you can climb up about 10 feet on breakdown to an upper room with a high ceiling, about 60 feet high. Air appears to be coming from above.

At the intersection, another passage goes east for about 300 feet to a room with 40-foot-high domes. From there you can climb up about 30 feet to another room with chert balls on the floor called the Cannonball Room. From there a passage goes east, but quickly twists around to the west and ends after about 150 feet with a possible high lead above. Another passage leaves the room to the west and goes about 90 feet and ends with air coming from above.

The stream passage intersected by the pilot bore at Large Tunnel Cave continues east for about 270 feet where the stream sumps in a 30-foot-high canyon passage. Mud was found all the way to the top of the canyon and it was speculated that the entire passage floods during high stream flow. The stream passage goes upstream from the intersection, in a westerly direction, for about 600 feet where the stream comes from a sump on the right side of the passage.

Exploration showed that these two caves did not connect to Gap Cave. Currently, they are gated and not open to the public. However, the high level air movement noted by Thomison may indicate an upper level in Small Tunnel Cave waiting to be explored.

Gap Cave Today

Gap Cave is currently open to the public. By 1992, the commercial section of the cave had been severely vandalized to the point that it had lost much of its attraction to visitors. This vandalism consisted of graffiti on the walls, litter on the ground, and the stairs and bridges for the commercial trail had deteriorated into poor condition. Much of this damage occurred after the cave was closed to tourists.

The right to operate commercial cave tours in Gap Cave was transferred to the National Park Service in 1992. Park Service staff and members of the Pine Mountain Grotto of the National Speleological Society have cleaned the cave of trash and graffiti and removed old

Figure 3.39. Bill Wilder crossing the stream in Gap Cave. Photo by Ken Storey, February 2007.

Figure 3.40. Jon Lillestolen and Richard Knapp at the Soldiers Cave Entrance to Gap Cave. Note gate and handrails along commercial trail. Photo by Cheryl Pratt, April 2004.

handrails, rock monuments, some of the asphalt trail, and the old electrical lighting system. Thanks to this restoration, the cave now closely resembles its original condition. The cave was reopened to the public in 2000. In 2002, the National Park Service removed the asphalt from old Route 25E, in an effort to restore the road to look like the path that had been used during Daniel Boone's day.

The current tour uses electric lanterns for light. With all electrical lighting gone, this gives the visitor a much more *wild cave* feeling during the tour. Visitors can now view large rooms and many beautiful cave formations. Gap Cave is once again a beautiful and interesting place to visit. The cave tour lasts 2 hours and is led by National Park Service rangers. The tour begins with an uphill half-mile walk from the Daniel Boone Parking Area to the cave entrance. The tour enters through the old Cudjos Cave Entrance and exits through the Soldiers Cave Entrance. Your visit will cover one mile of passages on four different levels. There are 183 steps in the cave, so visitors should be in reasonably good physical condition to take this tour. After exiting the cave it is a downhill half-mile walk back to the Daniel Boone Parking Area. Please note that children under the age of five are not permitted on cave tours.

At the present time, the schedule of tours is as follows:

January 1 – March 29
Weekends only at 10:00 A.M. and 2:00 P.M.

March 30 – May 24
Weekdays at 10:00 A.M.
Weekends at 10:00 A.M. and 2:00 P.M.

May 25 – September 1
Daily at 10:00 A.M. and 2:00 P.M.

September 2 – November 30
Weekdays at 10:00 A.M.
Weekends at 10:00 A.M. and 2:00 P.M.

Figure 3.41. Formations and rimstone pool. Photo by Cheryl Pratt, August 28, 2004.

The Park Office may be reached by telephone at this number:

606-248-2817

Cumberland Gap National Historical Park may be reached by e-mail by going to the following Web site:

http://www.nps.gov/cuga/contacts.htm

Click on the appropriate link to contact the Park Office by e-mail.

December 1 – December 31
Weekends only at 10:00 A.M. and 2:00 P.M.

This schedule is subject to change, so it is advisable to contact the Park well in advance of your trip.

Reservations for cave trips are recommended and can be made up to 30 days in advance by calling the telephone number given below. You may also contact the Park Office by writing to:

Cumberland Gap National Historical Park
PO Box 1848
Middlesboro KY 40965-1848

Directions

From Knoxville, go northwest on I-75 for 34 miles to the Caryville–Jacksboro Exit (Exit 134). Turn right on U.S. Highway 25W and proceed 8 miles to LaFollette, Tennessee. Highway25W turns left (north), but continue east on Highway 63 and proceed 32 miles to the intersection of U.S. Highway 25E. Turn left onto U.S. Highway 25E and proceed north for approximately 12 miles to the entrance to the Cumberland Gap Tunnel. This tunnel is 4,600 feet long. When you exit the tunnel you will be in Kentucky. The main visitor center for Cumberland Gap National Historical Park is 0.8 mile ahead and on the left.

The Visitor's Guide to Cumberland Gap National Historical Park contains an excellent map of the park, including a detailed map of the gap area itself. Using this map for directions to reach Gap Cave, go back through the Cumberland Gap Tunnel, turn left (east) on Highway 58, then take the second road to the left into the community of Cumberland Gap. Turn right into the Daniel Boone Visitor Information Center and park. Follow the Wilderness Road Trail left (west) for 0.3 mile to a fork. The right fork continues steeply uphill to the old Cudjos Cave Entrance. Immediately below, you will see a large stream of water emerging from the lower Stream Entrance. Just before you reach the Cudjos Cave Entrance, an even steeper trail branches off to the right and continues uphill to the Soldiers Cave Entrance. Both the Cudjos Cave Entrance and the Soldiers Cave Entrance are gated. The Stream Entrance becomes filled with water only a few hundred feet inside and cannot be used to enter the main cave.

Figure 3.42. Anne Elmore by commercial water pipes inside the Gap Cave Stream Entrance. Photo by Cheryl Pratt, May 2004.

Side Trips

Since Gap Cave is currently located inside the boundaries of Cumberland Gap National Historical Park, you will want to combine your visit to Gap Cave with a visit to the surface features of this park. The main feature, for which the park is named, is Cumberland Gap. This natural "gap," or natural break, in the Appalachian Mountains was the path followed by the early explorers and settlers into what was then the wilderness. It is estimated that between 1775 and 1810 that as many as 300,000 settlers moved west through the gap and into Kentucky. This wilderness later became the states of Kentucky and Tennessee. The park consists of 22,000 acres and is very rugged and heavily forested. This area is the junction of Kentucky, Tennessee, and Virginia, and the park includes portions of all three states.

One very interesting feature of the park is the Cumberland Gap Tunnel that carries U.S. Highway 25E northwest from Tennessee **under** Cumberland Mountain and into

Figure 3.43. Cheryl Pratt by rimstone dams in Gap Cave. Photo by Ken Storey, February 2004.

Kentucky. This modern, four-lane tunnel was begun in 1980 and was not completed until October 1996. The total cost of constructing the tunnel was $280,000,000. Before the tunnel was built, Highway 25E wound tortuously over Cumberland Mountain and was a major bottleneck in the highway system. The tunnel itself is a modern engineering marvel and is 4,600 feet long. During the construction of the tunnel, two new caves were discovered. Neither cave had a natural entrance to the surface so they had never been explored previously. Once the tunnel was completed in 1996, the Park Service restored old Highway 25E to how it would have looked as an early 19th century wagon road.

The visitor center and park headquarters is located in the Kentucky portion of the park, on Highway 25E and west of the Cumberland Gap Tunnel. This is the ideal place to begin your visit to the park and to learn what activities are available. You will need to purchase your cave tickets here at the visitor center. The cave is open year-round, but the tour schedule varies with the season.

Another natural feature in the park that you may wish to visit is **Sand Cave**. Sand Cave is not a true cave, but a large rock shelter that consists of a 75-foot-high sandstone overhang. Sand Cave is located near the eastern end of the Park and can be reached only by an 8-mile round trip hike. Sand Cave is named for the soft sand floor that covers the ground beneath the overhang.

There is a paved road (known as Skyland Road) that climbs the east side of Cumberland Gap to an overlook known as **The Pinnacle.** From here you will have a wonderful view of Cumberland Gap and the surrounding countryside. This view is particularly nice when the surrounding hardwood forest is at the peak of its fall colors.

Chapter 4

Indian Cave

The Discovery of Indian Cave

American Indians were accomplished cave explorers. Almost every cave in Tennessee with a reasonably large entrance shows some traces of exploration by these prehistoric people. According to cave historian Jim Whidby:

> It is well known that Indian Cave was visited by early American Indians. It certainly afforded a convenient and perfect shelter as well as an ideal water source, since the cave's stream flows out of the cave and into the Holston River. A Cherokee village is said to have existed just west of the main cave entrance. Ancient Indian mounds were once located in the area before being vandalized years ago. Local residents tell of a Frenchman who used the cave entrance when in the area trading with the Indians. During the winter of 1761, a peace agreement with the Cherokee Indians was made at Long Island (Kingsport, Tennessee). Lieutenant Henry Timberlake volunteered to go down the Holston River by canoe to meet the Cherokee Chiefs at Chota, to confirm the agreed-on peace with the Cherokee people. Timberlake took along with him a "serjeant" and an interpreter. The three spent two nights in a "large" cave 50 feet above, and overlooking the Holston River. This description is very possibly Indian Cave, which could have been easily seen by travelers from the river. Both nights they tell of being kept awake by the "howling of wild beasts in the cave." The sounds they heard have been speculated to have been bears which were known to have used Indian Cave long ago.[1]

The East Tennessee Grotto of the National Speleological Society began exploring and mapping caves in this area in 1982. It was on a surveying trip in Indian Cave in 1983 that they made the following discovery:

> On July 4, 1983, Jeff Bowers and Mike Whidby were surveying a low stream crawl passage. Jeff noticed some bones in a mud bank. A tooth was removed and later taken to the University of Tennessee Anthropology Department. The tooth was examined by Dr Pat Wiley and Dr Charles Faulkner and identified as being human. On July 7, 1983, Jeff returned to the cave along with Jim Whidby, Brian Ball, and Mike Green. Bone material and a bi-face flint knife were found within a 6-square-foot area in the dough-like mud. Several hours were spent in the low crawlway. Jim Whidby took many photographs of the entire area and material collected. Later Dr Wiley determined that the remains were that of a "robust" male American Indian, approximately 25 years of age at the time of death which he estimated occurred between 3,000 to 5,000 years ago. This time period was based on the style of bi-face knife found. Indian Cave was assigned Tennessee Archaeological

---

1 Jim Whidby, "Indian Cave Park, Grainger County, Tennessee: A Premier American Show Cave—Lost In Time," *SERA 2004 Guidebook*, pp 6–16.

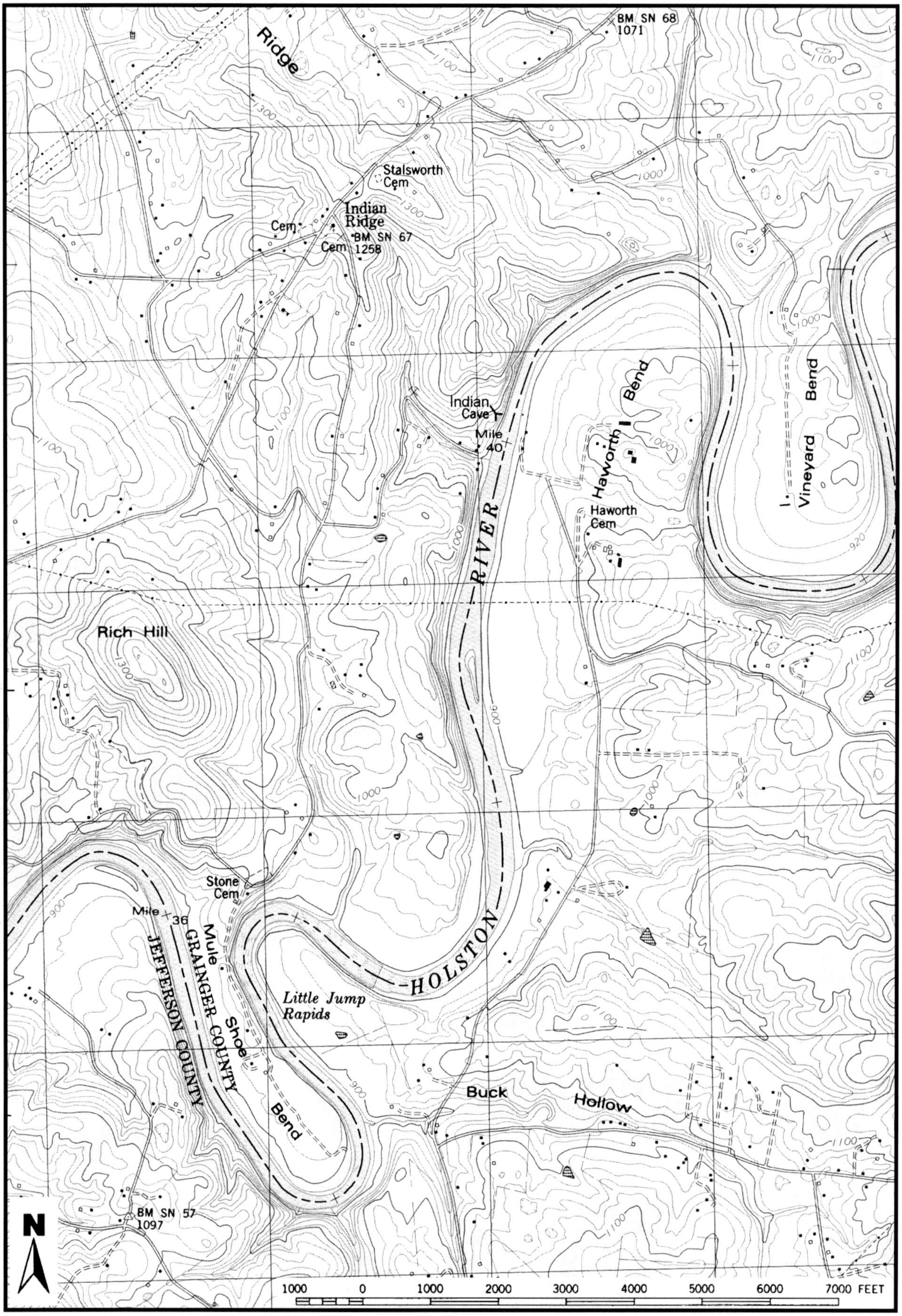

Figure 4.1. Topographic map showing the location of the entrance to Indian Cave.

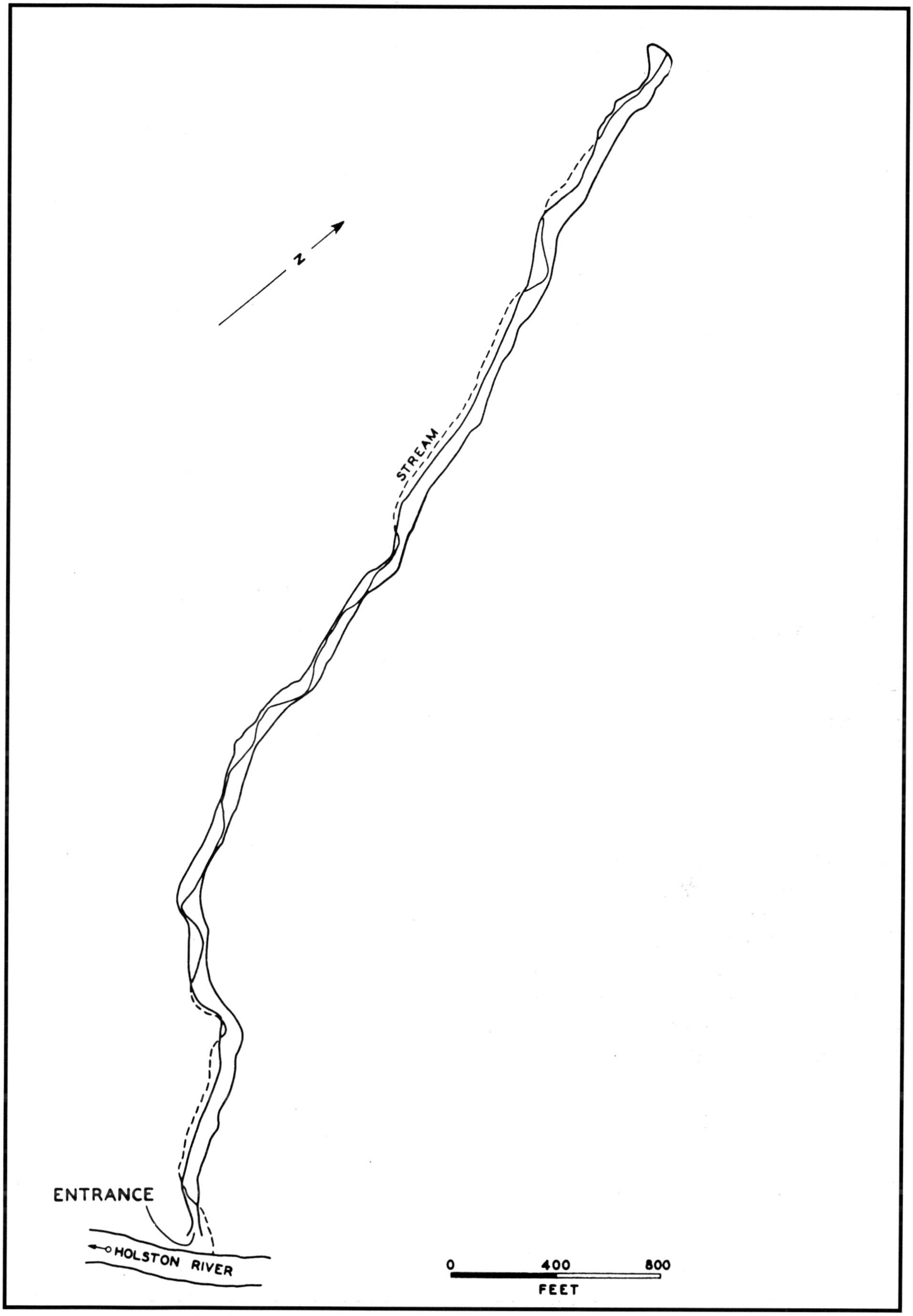

Figure 4.2. Map of Indian Cave from Caves of Tennessee, *page 214. Berlen C. Moneymaker, 1929.*

Site # 40GR14.[2]

American Indians in this area explored caves with cane torches. Cane was a very abundant plant in this area before white settlers cleared the canebrakes and converted them into fields for farming. On October 1, 1985, Dr Charles Faulkner, George Crothers, Charles Hall, and Jim Whidby collected cane torch material from three areas within Indian Cave. These samples were sent to the Smithsonian Institution Environmental Research Center in Rockville, Maryland, for C-14 dating. The results of the three dated samples are AD 255, AD 1290, and AD 1550.[3]

Early Caucasian hunters and trappers often used cave entrances for shelter. The early settlers investigated caves as a possible source of minerals. Whidby provides additional information of the early history of the cave:

> Other legends allude to a massacre of white men which reportedly occurred on the Holston River in front of the cave in the 1700s. This story further states that some of the people killed had smallpox, resulting in an epidemic among the Indian tribe.
>
> Folklore speculates that James Donelson and his party may have visited the cave about 1779 while enroute down the Holston River on their way to Middle Tennessee where they settled Nashville.
>
> In the 1800s a grist mill stood outside the cave entrance and was powered by the water flowing from the cave. "Uncle" Jimmy Reynolds operated the mill and the "Reynolds Ferry" at the cave prior to 1900. A concreted waterway carried water from the mouth of the cave to the mill and is all that now remains of the early mill.[4]

Other sources report that a French trader used the cave's entrance as a trading post in the early 1700s. He is supposed to have traded with the Cherokee Indian village that was located just west of the entrance.

In 1869, the family of North Carolina Governor Hoke Smith purchased the cave and mined bat guano. Bat guano was a valuable natural fertilizer.

Figure 4.3. Photograph of Major General Robert Fredrick Hoke. From the Jim Whidby collection.

The History of Indian Cave

Jim Whidby provides the following history of Indian Cave from the time the area was first settled by pioneers' families:

> Ownership of the cave property can be traced back to W.B. Mitchell, who transferred a hundred acres on July 21, 1869, to former Confederate Major General Robert F. Hoke (1837–1912) of North Carolina. On October 29, 1913, the property was

2 Jim Whidby, "Indian Cave, Grainger County, Tennessee & The East Tennessee Grotto," *The Speleotype*, vol 39, no. 2, pp 78–80.

3 Ibid.

4 Jim Whidby, "Indian Cave Park, Grainger County, Tennessee: A Premier American Show Cave—Lost In Time," *SERA 2004 Guidebook*, pp 6–16.

transferred to Van Wyck Hoke from Lydia A. Hoke. During the ownership by the Hoke family, General Hoke shipped large amounts of bat guano to Atlanta, Georgia, for fertilizer. It was said that General Hoke planned to open the cave to the public, but this did not happen during his ownership. Before the cave was commercialized it was a major grey bat (*Myotis grisescens*) maternity cave with large amounts of bat guano in several areas of the cave. It is said that during early commercialization, workers "torched the bats from the cave." It has been estimated by the U.S. Fish and Wildlife Service that in excess of 100,000 gray bats once used the cave as a maternity roost. In addition to the gray bats, the cave has also been home to the eastern pipistrelle bats (*Pipistrellus subflavus*) and the big brown bats (*Eptesicus fuscus*).

On January 4, 1916, the Indian Cave Park Association was formed, and a hundred shares of stock were issued for sale at $50 per share. The officers of the corporation consisted of five directors, which included a vice-president, a secretary, and a treasurer. The incorporators were R.M. McConnell, Charles H. Brown, R.F. Colgan, Charles M. Seymour, and Howard Cornick.

The incorporators first meeting was held 12 days later at the law offices of Cornick, Franz, McConnell, and Seymour in the East Tennessee National Bank Building in Knoxville. McConnell chaired the meeting and Howard Cornick acted as secretary. The first order of business was the presentation of the charter incorporating the cave park. At this meeting all directors except Cornick stepped down to allow the following new directors to be accepted

Incorporated under the laws of the State of Tennessee

Number 13

Shares

Indian Cave Park Association

Knoxville, Tennessee

This Certifies that _______ is the owner of _______ Shares of the Capital Stock of

Indian Cave Park Association

transferable only on the books of the Corporation by the holder hereof in person or by Attorney upon surrender of this Certificate properly endorsed.

In Witness Whereof, the said Corporation has caused this Certificate to be signed by its duly authorized officers and to be sealed with the Seal of the Corporation this _______ day of _______ A.D. 19__

Attest

Secretary

President

Shares $50.00 Each

Figure 4.4. $50.00 stock certificate for Indian Cave Park Association dated 1916. From the Jim Whidby collection.

Figure 4.5. Photograph of E.E. Wooten and Hunter Chapman about 1930. From the Jim Whidby collection.

the cave property. In a 1926 article it was claimed that the cave had the "longest stalagmite" in the world. In 1928 a brochure of the Smoky Mountains area described the cave with 27 photos of the cave formations and the property. It further stated that Indian Cave was one of the most interesting spots adjacent to the mountains, and that roads leading to the cave were graded and more suitable to travel by automobiles.[5]

According to Whidby, this initial commercialization consisted of rather minor improvements. However, in 1930, Mr Hunter B. Chapman became the new president of the corporation and the capital stock was increased from $5,000 to $150,000. Whidby gives the following details:

unanimously: M.M. Whittle, W.O. Whittle, E.E. Wooten, and H.M. Johnson. Stock was sold to each in the amount of $250.00 for five shares and, in addition, Wooten purchased another $3,000, making him the majority stockholder.

Commercialization work began, but the cave did not officially open to the public until May 30, 1924. Transportation from New Market was offered by carriage and the Indian Cave Ferry transported these visitors across the Holston River to

> Chapman was the owner of Shenandoah Caverns in Virginia, which had opened commercially less than ten years before. Shenandoah had already proven to be a great success with more than 87,000 visitors per year. Chapman had learned of Indian Cave from a railroad

5 Ibid.

Figure 4.6. Photograph of the entrance to Indian Cave in 1931. From the Jim Whidby collection.

Figure 4.7. Photograph of the Indian Cave Ferry about 1940. From the Jim Whidby collection.

official while on a trip to Washington, D.C., After learning of the cave's beauty, he became convinced that Indian Cave could become another Shenandoah. He, in part, seemed to be attracted due to some similarities in the two caves, such as both being near the same U.S. Highway 11 and both caves near towns named New Market. He also was attracted by the legends of the Indians' use of the cave and grounds.

According to Hunter Chapman's daughter-in-law, Louise, he spent about $32,000 to become the majority stockholder, and then over $79,000 on internal improvements in the cave, including better trails. He replaced the old wooden bridges in the cave with 14 concrete bridges. These concrete bridges remain in the cave today, most in their original condition. Additional improvements included more than 10 miles of lead-shielded electric cables which was installed the entire 5,150-foot length of the cave. A dam was built at the end of the cave, creating a lake for boat rides. The remains of a flat bottom boat can still be seen on the mud bank, moored to its post.

By 1931 Indian Cave was advertised as the largest electrically illuminated cave in the world, and two years later 100,000 folders about the cave were distributed over the nation. More than 1,800 electric bulbs, ranging from 200 to 500 watts, would light the cave's "borehole" passage. Spectacular colored light shows became a major attraction. Twenty-four hundred volts of electricity was furnished to the cave feeding several large transformers. All but one of these transformers are still inside the cave. The earliest power source was a Ford tractor engine installed outside of the cave on the hillside to produce the first electricity for the cave.

1930 and 1931 saw the greatest improvements and commercialization of the cave. Hunter Chapman's son, Paul, became manager, and Paul's wife, Louise, became executive secretary. Part of the year

Figure 4.8. Photograph of Indian Cave Park in the 1930s. From the Jim Whidby collection.

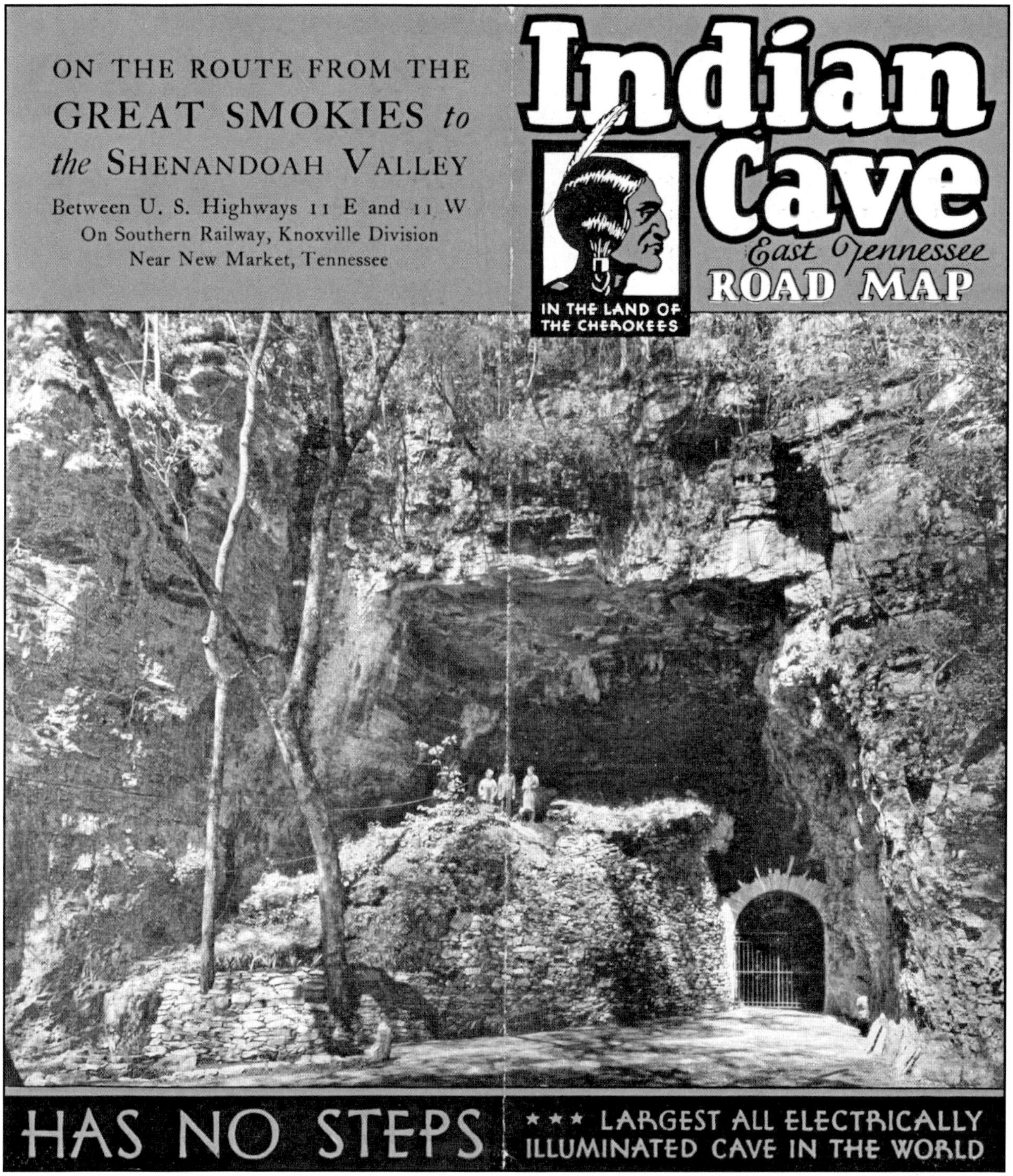

Figure 4.9. Front and back of an old brochure for Indian Cave. From the Speleo Archives of Gary K. Soule.

they lived in a screened, mostly chestnut wood cottage near the cave entrance, where they also ran a small general store. Water for them and their guests was lifted from the stream exiting the cave to a reservoir on the hillside by a ram pump. They also maintained eight cabins, of which only two are still standing. They also built a camping area on the hillside above the cave. These cabins in 1939 rented for from seven to twenty dollars a week. The cave park was only operated in the summer months. During the winter the Chapmans moved their residence back to Shenandoah Caverns.

Most of the visitors to the park came

A VISTA OF OLD PEKIN

PASS AT THERMOPYLÆ

GROVE OF THE GNOMES

PILLARS OF HERCULES

IT is universally acknowledged that the Appalachian area of the East is the scenic section of America, offering the tourist natural attractions and a picturesqueness possible only where towering mountains and turbulent streams are the background of changing vistas. Particularly is this true in the Southern Appalachians, of which the burly Blue Ridge and the Great Smokies are integral parts.

It is in this section that the great caverns of America have been discovered and the necessary development completed to make them accessible to the never-ceasing stream of visitors who are seeking the unusual and interesting along the motor highways of the historic South.

For four hundred years INDIAN CAVE, in East Tennessee, has been known, in tradition and in fact, as one of the outstanding subsurface formations in America. The tradition has been handed down by generations of Cherokees who found a haven of safety in the immense spaces beneath towering cliffs which flank the Holston River, and until this day evidences of the Red Man's habitation may be seen on the smoked strata of the cave.

While for over a century the early settlers of East Tennessee knew of and visited the immense cavern, only within recent years has it been the objective of tourists and now, at great expense, Indian Cave has been electrified, and its vastness is verified by the fact that many miles of leaded cable and over eighteen hundred bulbs of from two to five hundred watts are required for its electrification.

The illumination revealed not only a cave of immense underground reaches, but one of surpassing beauty, for in the travertine deposits nature had created countless thousands of stalactites and stalagmites, great masses of drip stone and flow stone and translucent draperies, woven in the loom of Time, in tinted warp and glistening woof. It may be safely asserted that no caverns in America, easily accessible, have within their depths as many formations of equal diversity, size and shape as may be found in INDIAN CAVE.

Within a few yards of the great arched entrance to the Cave flows the beautiful Holston River, at the base of the mighty palisades which extend for miles on either side of the entrance and, with the Great Smokies in the distance, one finds here a picture without parallel in the Southern Blue Ridge area.

INDIAN CAVE is but twenty-four miles from Knoxville, between U. S. 11 E and 11 W. Leave U. S. 11 E (Andrew Johnson Highway) at New Market or U. S. 11 W (LEE HIGHWAY) at Blaine.

If on Dixie Highway use Route 61 just east of Maynardville to Blaine on Lee Highway.

It is 100 miles from Bristol, 100 miles from Asheville, 145 miles from Chattanooga and 24 miles from Knoxville.

Your visit to the Great Smokies or East Tennessee is not complete without a visit to INDIAN CAVE.

Arrangements will be made with Southern Railway for stop-over privileges at New Market, Tennessee, on all tourist tickets, where transfer to Caverns will be made. For information, maps or folders write

INDIAN CAVE PARK ASSOCIATION
NEW MARKET, TENNESSEE

or

SHENANDOAH CAVERNS CORPORATION
SHENANDOAH CAVERNS, VIRGINIA

Figure 4.10. Inside pages of an old brochure for Indian Cave. From the Speleo Archives of Gary K. Soule.

by way of New Market, which was located about 5 miles away on the opposite side of the river. The Indian Cave Ferry was very important as an access to the cave. Roy Vineyard took over the Indian Cave Ferry operations in the 1920s, prior to the Chapmans' arrival. Roy was still running the ferry when the Chapmans left the area in the 1940s. The ferry was constructed mostly of wood with a steel frame and was carried from one side of the river to the other side by the river current. A large "ship's wheel" was used to apply tension to the cable which was attached to the main cable that spanned the river. Applying tension to the cable would cause the ferry

Figure 4.11. Louise Holland Chapman at the "ship's wheel" that controlled the motion of the Indian Cave Ferry, about 1930. From the Jim Whidby collection.

Considering the relative remoteness of Indian Cave, the fact that it was over 5 miles off the main highway, and that the ferry could accommodate only four cars at a time, it seems amazing that the cave was a commercial success. However, one must consider that Highway 11 was one of the most important highways in the country at that time, linking the area of Washington, D.C., southwest to Knoxville, Chattanooga, Birmingham, and New Orleans. In its day, Highway 11 was as important as any interstate highway that exists today. The fact that the owner printed 100,000 brochures for the cave shows that it was extremely well advertised for its time. And Shenandoah Caverns, which was operated by the same owner, is still one of the most visited show caves in the United States.

to move sideways in order to take advantage of the river current. The ferry could accommodate up to four automobiles at a fee of 25 cents each. In the early 1950s the ferry sank on the west side in the river; a part of the frame can still be seen during very low water times. It is said that Roy continued to transport people across the river in a small boat. Indian Cave boomed throughout the 1930s.[6]

Jim Whidby gives more details of the early operation of the cave:

> Admission to the cave during the 1930s was one dollar per person. An annual 4th of July celebration was held at the park each year since 1925. By 1938 attendance was estimated at 5,000 according to local newspapers. Visitors from eight states

6 Ibid.

Figure 4.12. An old postcard showing "A Scene in Indian Cave, East Tennessee." From the Speleo Archives of Gary K. Soule.

were believed to be in attendance, to hear speeches by well-know people including many politicians including the Governor of Arizona, R.T. Jones, who had been born in the county. Lunches, cold drinks, and ice cream were sold, visitors were entertained by baseball games, and about 30 American Indians in their native costume were hired to perform ceremonies.[7]

The cave continued in operation through the 1940s and into the 1950s and 1960s:

7 Jim Whidby, "Indian Cave, Grainger County, Tennessee & The East Tennessee Grotto," *The Speleotype*, vol 39, no. 2, pp 78–80.

Figure 4.13. An old postcard showing the "Tower of Babel, Indian Cave, East Tennessee." From the Speleo Archives of Gary K. Soule.

Figure 4.14. An old postcard showing a bridge across the cave stream in Indian Cave. From the Speleo Archives of Gary K. Soule.

Indian Cave remained in the Chapman family until 1947 at which time it was purchased by William Eugene "Gene" Monday, a Knoxville realtor. Monday purchased 132 acres including the cave, which was claimed to be the third largest commercial cave in the United States. Soon Monday began construction of a lake which would provide swimming, boating, and fishing. Water from the cave was piped underground to the lake. In 1948, Monday hired Charlie and George Norton to manage the cave and they continued until 1968. In 1960 admission to the cave was $1.50. In the early 1960s Indian Cave became an official fallout

Figure 4.15. An old postcard showing some of the original lighting system in Indian Cave. From the Speleo Archives of Gary K. Soule.

1925 **1939**

14th Annual Home Coming

JULY 4th CELEBRATION

Home Coming and Reunion of Families and Friends of Grainger, Jefferson and Other Counties at

INDIAN CAVE

Greetings to the thousands who have been attending each year—GIVING YOU A BETTER PROGRAM than ever before. Bring the friends with your lunches (or get sandwiches and drinks on the grounds).

PROGRAM

OPENS PROMPTLY AT 10:00 A. M.

	Music directed by Hon. W. I. Daniel, Clerk of the County Court, Rutledge
	Welcome Address—City Atty. James Iddins, Jefferson City
	Response—Dr. W. E. Gallion, Mayor, Oakdale
10:45	Literary Gems—County Historian C. C. Justus, Knoxville
11:05 to 11:45	Address—Hon. Burgin E. Dossett, Manager Social Security Board, Knoxville
12:00 to 1:30	Noon Intermission
1:30 to 2:10	Patriotic Address—Hon. Hamilton S. Burnette, Judge of Knox County Circuit Court
2:10	Address—Notable Contacts and Values in the Tennessee Senate and House—Judge S. H. Justice

INDIAN TRIBE HERE About 30 Indians, costumed, with ceremonies in Cave. Two baseball games will be held during the day, the first game starting at 9 a. m., in the Jefferson and Grainger County League.

Figure 4.16. Handbill advertising the July 4, 1939, July 4th Celebration at Indian Cave. From the Jim Whidby collection.

shelter. According to the yellow and black shelter sign, 9,065 people could be sheltered in the cave. Also at this time a two-story motel was built, but just before completion, an electrical fire destroyed part of the structure. The lower level of the building was saved and completed and has been used to this day for living quarters, a restaurant, a museum, gospel concerts, and wrestling events.[8]

Eventually, Indian Cave fell on hard times, as described here by Jim Whidby:

> The decline of this very popular cavern and park seems to have begun in the 1950s. Two factors that most affected the park was the completion of Cherokee Dam in 1942. The dam is 4 miles up river from Indian Cave. In time the dam caused the demise of the ferry, as it was dependant on the water level of the Holston River. When the water was low Roy would transport people by a small boat. Without the ferry to transport people across the river, the cave could not experience the popularity it had seen in earlier years.
>
> The second factor that had a major effect was the completion of Interstate 40 on the opposite side of the river from the cave. The interstate highway took a large amount of north-south traffic from U.S. Highway 11W, located on the cave side, to the opposite side of the Holston River.[9]

In 1982 Daniel Osborne leased the cave from Gene Monday and began an effort to re-develop the cave. In October 1982 a Haunted Cave party was held. On May 8, 1984, the cave experienced an enormous flood during a period of heavy rain. Most of the cave trails were coated in mud and several of the concrete bridges were damaged.

8 Ibid.

9 Ibid.

Figure 4.17. Photograph of the chestnut wood cottage, Paul and Louise Chapman's summer cabin. Photo taken about 1930. From the Jim Whidby collection.

Figure 4.18. Photograph of the Fallout Shelter sign at the entrance to Indian Cave in the 1960s. From the Jim Whidby collection.

In 1989 Richard and Betty Dykes took over management of Indian Cave Park and continue to manage the cave to this day.

A Description of the Cave

The book, *Caves of Tennessee,* which was written by Thomas C. Barr Jr in 1961, gives the following written description of the cave:

> Indian Cave has been developed commercially and is electrically lighted. According to B.C. Moneymaker (unpublished thesis, 1929), it was opened to the public May 30, 1924, and was developed by the Indian Cave Park Association, although it had been known for many years prior to this date. A Cherokee village is said to have existed just west of the mouth.
>
> A metal gate bars the 30-foot arch of the main entrance, which opens in a bluff along the river. The cave stream issues from a lower opening. Indian Cave essentially consists of a single gallery 4,865 feet in length, 40 to 135 feet wide (average width 56 feet), and 5 to 50 feet (average 21 feet) high. (Moneymaker, 1929). The cave extends northwestward and is under the control of two perpendicular joint sets. Indian River runs through the cave and is bridged at several places by walkways. Dripstone decorations are numerous but occur sporadically and are not especially large.[10]

Indian Cave is the longest known cave in Grainger County, with a surveyed length of 2.58 miles (4.15 kilometers). This makes Indian Cave the 62nd longest known cave in Tennessee, out of over 8,000 known caves.[11] Indian Cave is number 344 on the USA Long Cave List.[12]

Like most large caves, Indian Cave has been surveyed several times. Moneymaker made the first survey in 1929. Harry Moore made the next survey in 1979. Jeff Bowers made the last, and most recent survey. The cave consists of a 5,150-foot-long main passage with many smaller side passages. The cave stream, known as the Indian River, flows the entire length of the main passage. The cave has three known entrances, all within 300 feet of each other. They are known as the Indian Cave Entrance, the Blowing Cave Entrance, and the Potato Cave Entrance.

Indian Cave Today

Indian Cave is still open to the public at the present time and is currently operated by Richard and Betty Dykes. The cave is open during the spring, summer, and fall, but is closed during the winter. Since Indian Cave is somewhat off the beaten path, you may wish to call their telephone number before you go to be sure the cave will be open on the day you plan

---

10 Thomas C. Barr Jr, *Caves of Tennessee*, Bulletin 64, Tennessee Division of Geology, 1961, pp 213–214.

11 Gerald Moni, *Tennessee Long Caves List*, July 31, 2007.

12 Bob Gulden, *USA Long Cave List*, April 16, 2007.

to visit. The telephone number is:

865-828-8047

The mailing address for Indian Cave Village is:

Indian Cave Village
1111 Indian Cave Rd
Blaine TN 37709-5823

Other activities that are currently held at the cave on a regular basis include wrestling events and a Haunted Cave around Halloween.

Since this cave is 7 miles off a main highway, it is somewhat of an adventure driving to the cave. However, once you arrive you will be amply rewarded with the unique beauty and historical surroundings of the place. Visiting Indian Cave is truly an adventure and is a pleasant change of pace from your visit to a more traditional, modern commercial cave.

Figure 4.19. Photograph of a sign advertising Indian Cave in downtown Knoxville prior to the 1982 World's Fair. From the Jim Whidby collection.

Directions

Indian Cave is located in Grainger County approximately 25 miles northeast of Knoxville. It takes approximately an hour to reach the cave from Knoxville. Below, are specific directions for locating the cave:

From Knoxville, drive east on I-40 for 5 miles to Exit 392. Take this exit and go northeast on U.S. Highway 11W. Follow 11W for 16.4 miles to the community of Blaine and turn right onto Fennell Road. Go 2.8 miles to an intersection. Turn right and continue on Fennell Road for 0.1 mile to an intersection. Turn left and continue on Fennell Road for 0.9 mile to an intersection. Turn right and continue on Fennell Road for 0.7 mile to an intersection.

Figure 4.20. Close up photograph of a biface stone knife discovered in Indian Cave on July 7,1983. From the Jim Whidby collection.

Figure 4.21. Photograph of Jim Whidby holding a biface stone knife at the time of discovery, July 7, 1983. From the Jim Whidby collection.

Turn right onto New Corinth Road and proceed 1.7 miles to the end of the road. New Corinth Road ends at the Holston River and the Indian Cave property. Indian Cave is located approximately 700 feet upstream, in a cliff overlooking the Holston River.

Indian Cave is a famous local landmark. Once you turn off U.S. Highway 11W onto Fennell Road, any local resident can give you directions to the cave.

Side Trips

Rather than return to Knoxville the way you came, when you return to Highway 11W, turn right (northeast) and proceed for 25 miles through rural, scenic Poor Valley. On your left you will be paralleling Clinch Mountain. When you reach the community of Bean Station, turn right on Highway 25E and proceed south for 15 miles to Interstate 81 (Exit 8). Turn right (west-southwest) onto I-81 and travel 8 miles to the junction with I-40. Go west on I-40 for 30 miles to Knoxville. Although this is rather a long loop (53 miles), you will be rewarded with a drive through some very scenic countryside.

Figure 4.22. A button or badge advertising Indian Cave Park. From the Speleo Archives of Gary K. Soule.

Figure 4.23. Photograph of Jim Whidby with charcoal and cane stoke marks left by prehistoric explorers in Indian Cave. Photo by Larry E. Matthews, January 10, 2008.

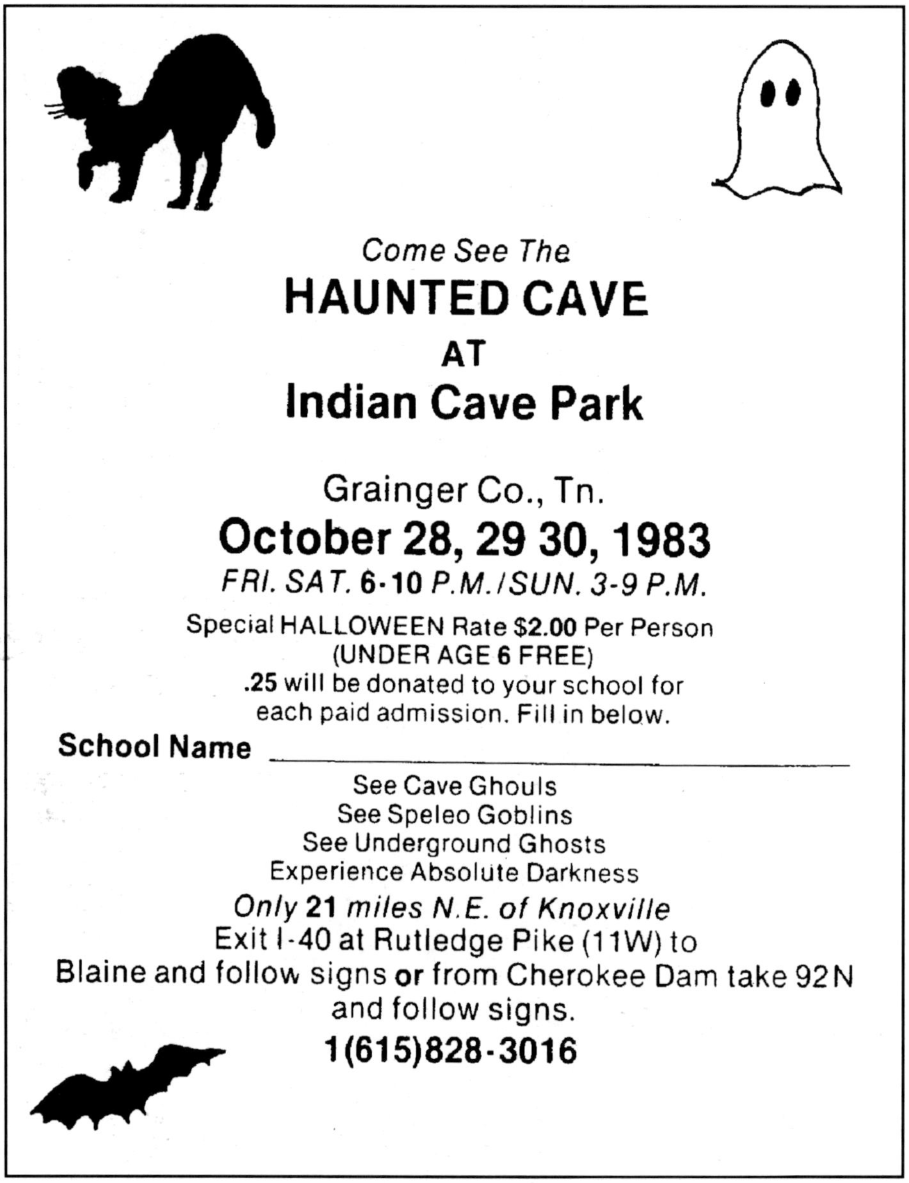

Figure 4.24. Flyer advertising the "Haunted Cave at Indian Cave Park," October 28, 29, 30, 1983.
From the Speleo Archives of Gary K. Soule.

Figure 4.25. Photograph of the current ticket office for Indian Cave Park. From the Jim Whidby collection, 1980s.

Figure 4.26. Photograph of the Blowing Cave Entrance. Photo by Larry E. Matthews. January 10, 2008.

Indian Cave

COMPLIMENTARY PASS
ISSUED TO:

THIS PASS ENTITLES THE BEARER
TO ONE FREE ADMISSION TO
INDIAN CAVE WHEN
ACCOMPAINED WITH
TWO OR MORE PAID
ADULT ADMISSIONS

DAN OSBORNE
MGR

WALKING TOURS ONLY

Figure 4.27. A pass for free admission to Indian Cave. From the Speleo Archives of Gary K. Soule.

Figure 4.28. Photograph of Richard Dykes, the current manager of Indian Cave Park. Photo by Jim Whidby, April 2004.

Chapter 5

Meredith Cave

The Discovery of Meredith Cave

Meredith Cave was known and used extensively by prehistoric peoples. William S. Webb's 1938 archaeological investigations in Meredith Cave revealed 13 burials near the entrance. Webb also reports ash and midden debris for a distance of 600 feet from the mouth and he uncovered some pottery and bone instruments.

Settlers began moving into what is now east Tennessee in significant numbers after 1775. Tennessee was granted statehood in 1796 and Campbell County, where Meredith Cave is located, was formed in 1806. It seems likely that these settlers discovered Meredith Cave sometime during the later part of the 1700s, but we will probably never know the exact details of the discovery.

An Archaeologist Visits Meredith Cave

William S. Webb conducted an archaeological study of the Norris Basin in eastern Tennessee and in 1938 published the results of that study as Bulletin 118 of the Smithsonian Institution's Bureau of American Ethnology. In that study, Webb describes 23 separate sites, which include villages, mounds, and caves. Below is his description of Meredith Cave:

Site No. 3.—SALTPETER CAVE

This cave, which was located on the land of Mr Stoke Meredith, was 11 miles east of LaFollette, in Campbell County, Tennessee. It was 2 miles west of Powell River and about 5 miles up the river from its mouth. About 50 yards from the entrance to the cave there was a spring which might have been used by the prehistoric inhabitants of the site.

The cave faces nearly due east. Its appearance was very different from that which it must have presented in prehistoric times. The bank of earth in front of the entrance was caused in part by the erosion of soil from the hillside above the cave mouth and in part by other factors. During the Civil War the Confederate Army used the cave as a source of saltpeter. Large piles of earth and ashes, the by-products of the process of extracting the niter from the soil, were carried out of the cave and dumped at the entrance. Excavation in the ash beds outside the cave yielded no evidence of prehistoric occupation.

The entrance to the cave was about 40 feet wide, with an overhead clearance of about 10 feet. From the entrance toward the rear of the cave, the floor descended and the roof ascended until, at a distance of 80 feet from the entrance, the ceiling was 40 feet high. The floor was covered with large rock masses, which had fallen from the roof. Beyond this large vault the cave was much smaller and fairly dry. At a point about 1,000 feet from the entrance there was a pool of clear water. The entrance to the cave will probably be above the high-water mark of Norris Lake, but, since the floor descends sharply, the rising waters will certainly fill the rear portion of the cavern.

Inside the entrance the sloping floor was covered with ashes and midden deposit to a depth varying from 3 to 6 feet for a distance of 80 feet. The midden material

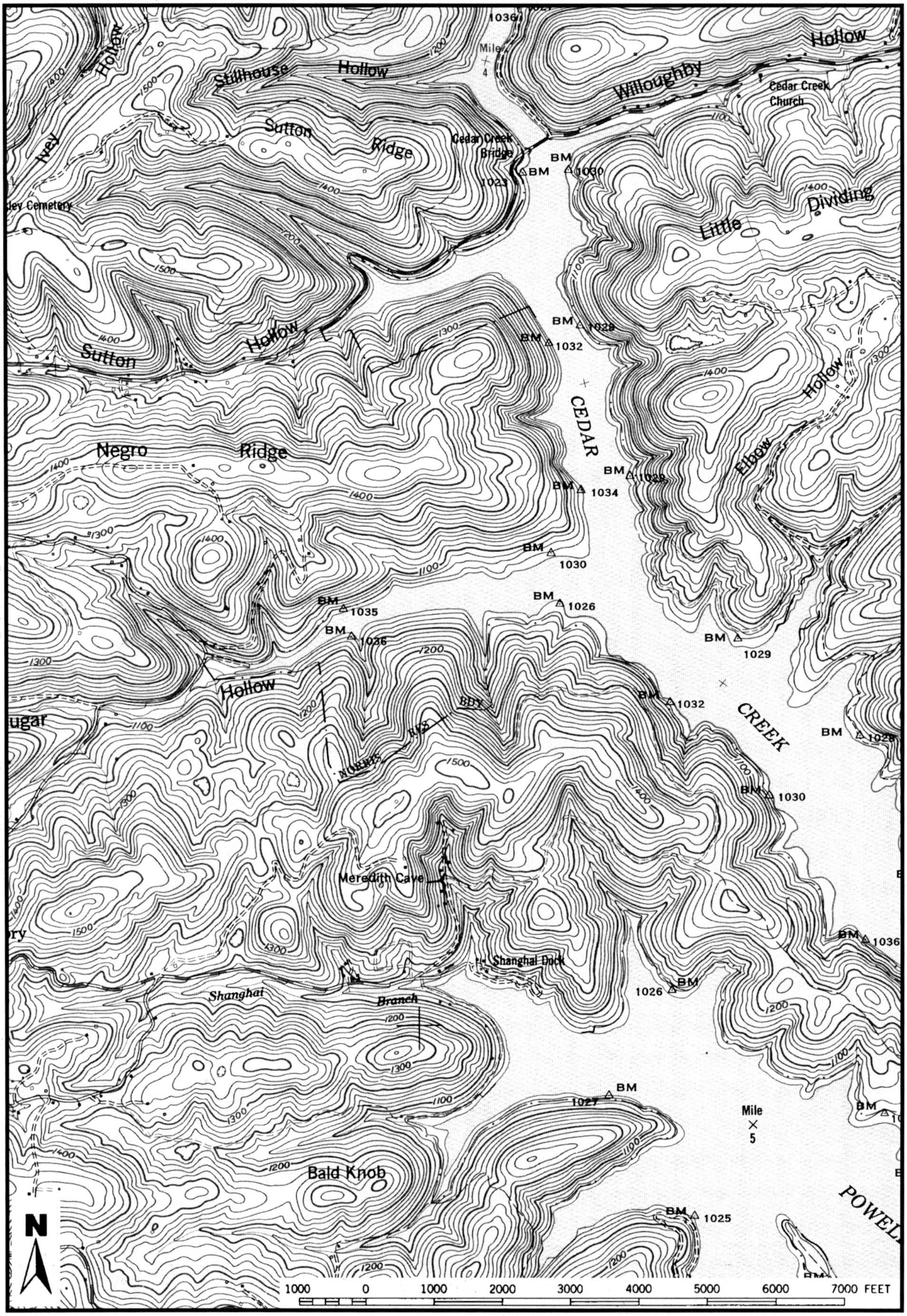

Figure 5.1. Topographic map showing the location of the entrance to Meredith Cave.

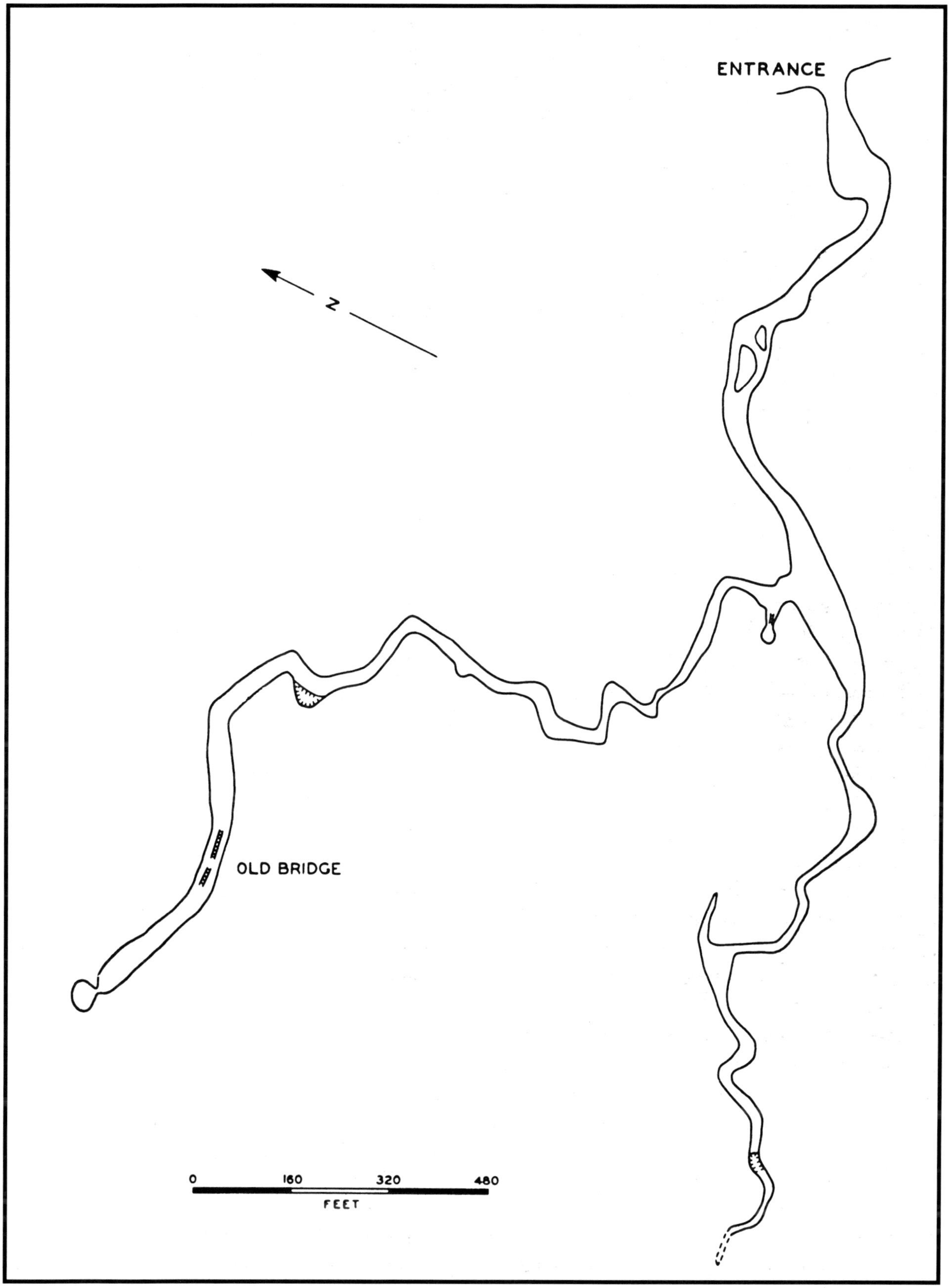

Figure 5.2. Map of Meredith Cave from Barr's Caves of Tennessee *(1961), page 90.*
Survey by Berlen C. Moneymaker, 1929.

Figure 5.3. Photograph of Berlen C. Moneymaker. Photo courtesy of Joan Moneymaker and Marcia Moneymaker.

contained much shell and many fragmentary animal bones. Beyond the bank of debris there seemed to be no further evidence of prehistoric occupation, although a careful inspection of the cave floor was made for a distance of 600 feet. A local inhabitant stated that some three years previously he had seen a burial wrapped in cane and bark resting on a ledge of rock in the cave wall, some 400 feet back from the entrance.[1] He offered to guide the party to it. Careful inspection failed to reveal its location, or to yield any evidence of occupation deep in the cave, except for the wooden vats used by the Confederates in leaching the soil for niter. It is believed that the information given by the man was correct, since this description checked with information obtained by exploration—a fact which he could not have known at the time. The burial he had seen had doubtless been removed by someone within the last three years.

Excavation of the cave floor was begun at a point 80 feet from the entrance where the midden deposit thinned out to meet the rock floor of the cavern. Here a trench was run from one side of the cave to the other. The earth was thrown behind the workers into the cave as the trench was exploited laterally toward the entrance. In this way the entire midden deposit in the cave entrance was examined. This resulted in the removal of a layer of ashes and midden material from 3 to 6 feet over an area of about 3,500 square feet.

Figure 7 is a plan of the cave entrance. It shows the location of the 13 burials and the fireplace which were uncovered. While the cave was as dry as most caves, the midden deposit at the entrance was damp and had long been subjected to the action of outside water. The water flowed into the cave as the result of the downward slope of the floor and because of the accumulation of earth and ashes described above which blocked the mouth of the cave. As a result of this condition, some of the skeletal material in the dampest part of the ash bed had decayed badly. Because of the frequent disturbances of the surface since prehistoric times, the data on the depth of the material discovered was considered to have little significance.

Burial No. 1.—The grave contained the skeletal remains of a child buried on its back with its head to the east. The legs were flexed on either side of the torso. The body had been wrapped in cloth and surrounded by the bark of trees. The covering was badly decayed, but some fragments were preserved.

Burial No. 2.—The grave contained the fully-flexed remains of a young female. The body had been buried with the head to the northeast. It had been covered with a large rock, which, in shifting slightly downhill, had caused the flexed leg bones to shift to the rear.

Burial No. 3.—In this grave were found the poorly preserved remains of an infant that had been buried with the head to the north. It was flexed and lay on the

1 The description of this burial matches burial goods associated with American Indian mummies found in Big Bone Cave, Tennessee.

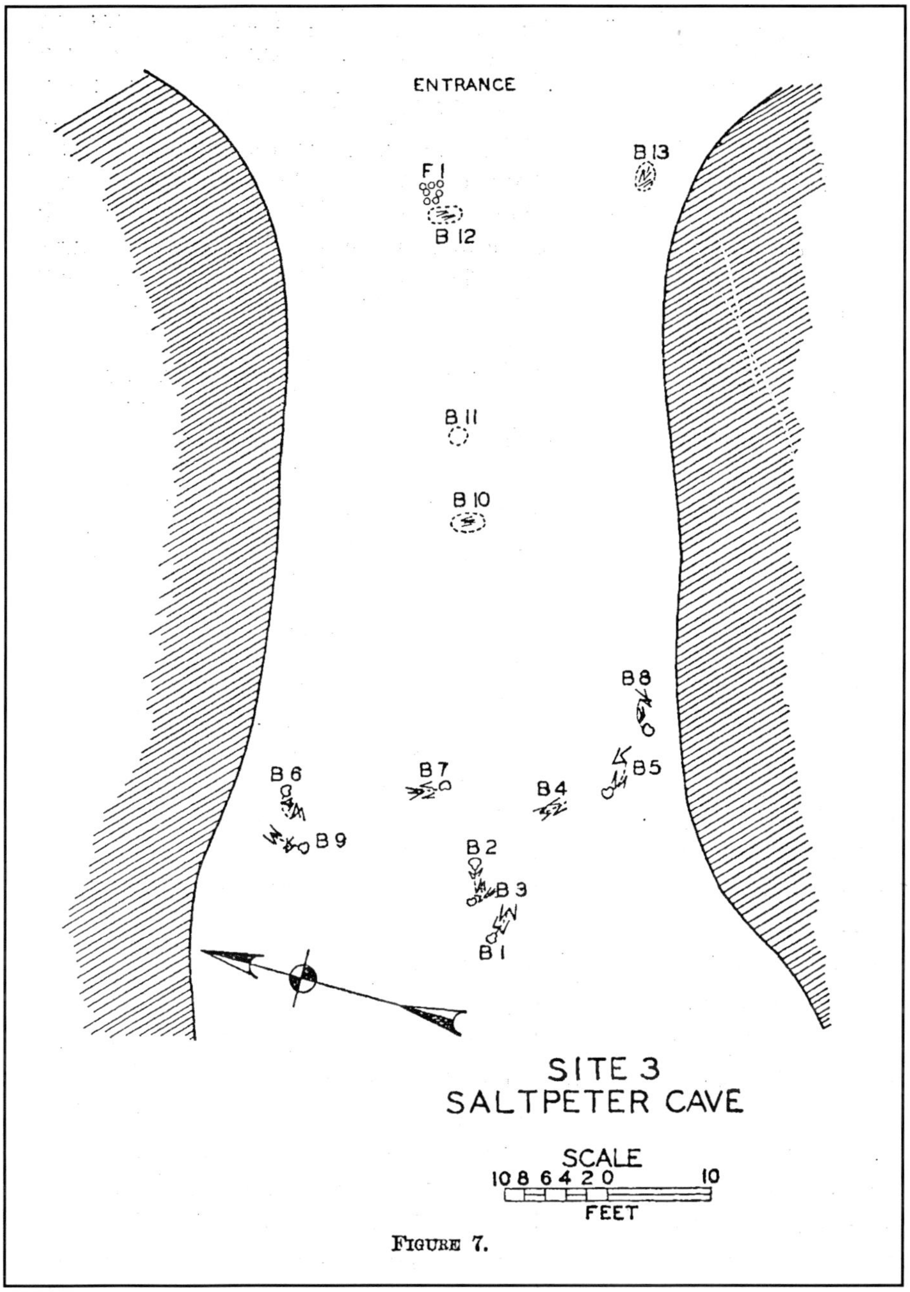

Figure 5.4. Map from Webb (1938) showing the location of the thirteen burials in the cave (figure 7).

left side. The skull had been crushed.

Burial No. 4.—This was a fully-flexed adult which had been placed on its stomach. The head was under the thorax; the left forearm lay at right angles to the humerus and rested under the torso. The knees were flexed and the feet were placed at the pelvis. The preservation of the skeleton was very good.

Burial No. 5.—This was the body of a child which had been buried on its back with its face turned to the left. The legs were flexed to the left. The femora were placed at right angles to the pelvis; the tibia and fibula were closely flexed. The skull was crushed.

Burial No. 6.—A fully-flexed adult, buried with the face down, was found in this grave. The knees were touching the chin. The pelvis bones and sacrum were missing. A bone awl was found between the feet and the skull.

Burial No. 7.—The grave contained the poorly-preserved skeleton of an infant. It was buried on the back with its head to the southeast. The skull was crushed and the leg bones were missing. There was a faint indication of a bark wrapping.

Burial No. 8.—This grave contained the partially-flexed remains of a child, buried on the right side. The skull was crushed. A dozen periwinkle shells were scattered on the body.

Burial No. 9.—This contained the partially-flexed skeleton of a child, buried on the right side, with the head to the south. The body had been covered with a large rock, as shown in plate 11, *a*. Plate 11, *b*, shows the same burial with the stone removed.

Burial No. 10.—This burial of an

Figure 5.5. Photo from Webb (1938) showing pottery fragments with various surface finishes from Meredith Cave (plate 13a).

adult was covered by a limestone slab 18 inches long and 12 inches wide. Many of the bones, including the skull and the pelvis, were missing. From the arrangement of the bones, shown in plate 12*a*, it would seem that this might be described as a flesh reburial. Some of the bones, the spinal column in particular, were still in anatomical order.

Burial No. 11.—This was a cremated burial. The body probably had been burned elsewhere and the residue of ashes, together with the artifacts, had later been deposited in the cave. The artifacts found included a drilled antler amulet, a double spatulate bone artifact—probably a weaving tool—three arrow heads, one bone awl, and a bone celt. The burial was covered by a large oval-shaped stone. When the stone was removed many bone fragments adhered to it, as shown in plate 12*b*.

Burial No. 12.—This was an adult burial covered with 10 large stones. The skull was missing, except for a part of the mandible. The bones were not in anatomical order. It appeared to have been a burial of disarticulated bones superimposed on a burned clay surface.

Burial No. 13.—The grave contained a pile of disarticulated bones of an adult, covered by a pile of stones. The skull was missing. A pestle was found in association with the bones.

Beside the burials the only other feature of importance encountered in the midden material was a stone hearth constructed of 25 stones in an oval arrangement, having a maximum length of 45 inches and a breadth of 33 inches. The fire pit adjacent to burial no. 12 was filled with ashes to a depth of 5 inches. It was 36 inches below the present surface level.

Pottery

The sherds from this site have a variety of surface finishes, as shown in plate 13*a*, but are alike in that all are hard gravel or sand tempered; the gravel tempering being of coarse texture. All rims are plain, and there are no lugs, handles, or other forms of decoration. Among the surface finishes the following may be listed:

1. Comb trailing, as shown on the lower right corner of plate 13*a*.
2. Cord-wrapped small cylindrical tool impression, as shown in the lower center of plate 13*a*.
3. Cord-wrapped paddling.
4. Textile impressed. This textile appears to be an unusual form of twined weave. The sherds are small and wholly different in appearance from so-called salt-pan pattern, as shown by the sherd second from the right in the upper line of plate 13*a*.
5. Grass-wrapped paddling, as shown by the sherd in the upper right-hand corner of plate 13*a*.
6. Stamped ware. The meshes are

square, 10 to the inch, and vary in size to rectangular meshes, 4 or 5 to the inch. By actual count the stamped ware was found to represent 25 percent of the total.

Plate 13*a* shows one sherd having a projection-like leg. The construction would seem to be the same as the four-legged vessels found on site no. 12.

Figure 5.6. Photo from Webb (1938) showing stone artifacts from Meredith Cave (plate 13b).

One fragment of a steatite vessel was taken from this cave.

Artifacts

Stone artifacts are shown in plate 13*b*. The flint implements are in general very crudely chipped. At the lower left is shown a "turkey tail" projectile point 4.8 inches long, very thin and beautifully made, which was found near the surface of this cave. For this reason it would seem that it was introduced into the cave at a relatively late period. A few perforated steatite disks and small celts were found, together with crude pestles, hammerstones, and mortars.

A seeming characteristic of this site was the use of crude two-hole gorgets of hematite, sandstone, or cannel coal, as demonstrated by the number of broken and unfinished specimens found.

Plate 14*a* shows samples of the awls taken from the general digging on this site. Awls were very numerous, being found by the hundreds. They were made from ulna of deer and wolf, deer scapula, split cannon bone of deer, bird bones, and deer horns.

Many other specialized bone artifacts were found. Plate 14*b* shows bone chisels, horn flaking tools, carved bone handles, fishhooks, hairpins, carved bear jaw, and cut bear femora, hollowed out and probably used to make a grease lamp, bone gorget and spoons made from shell, and the carapace of the box tortoise.

A double spatulate object of bone 10.75 inches long was taken from the deposited remains of a cremated burial. This object is highly polished and may have served as a weaver's tool.

Conclusions

The findings in this cave are in many ways similar to those reported from rock shelters in eastern Kentucky which have been attributed to members of the great Algonquian family. Among those characteristics which seem to be diagnostic the following may be mentioned:

1. Occupancy and burial under rock shelters and in the mouths of caves.
2. Predominance of flexed burials devoid of artifacts.
3. Occasional evidence of burials of dismembered bodies in the flesh and bundle burials of bones.
4. The occasional deposit in cave floor of partially cremated remains, the actual burning having taken place elsewhere.

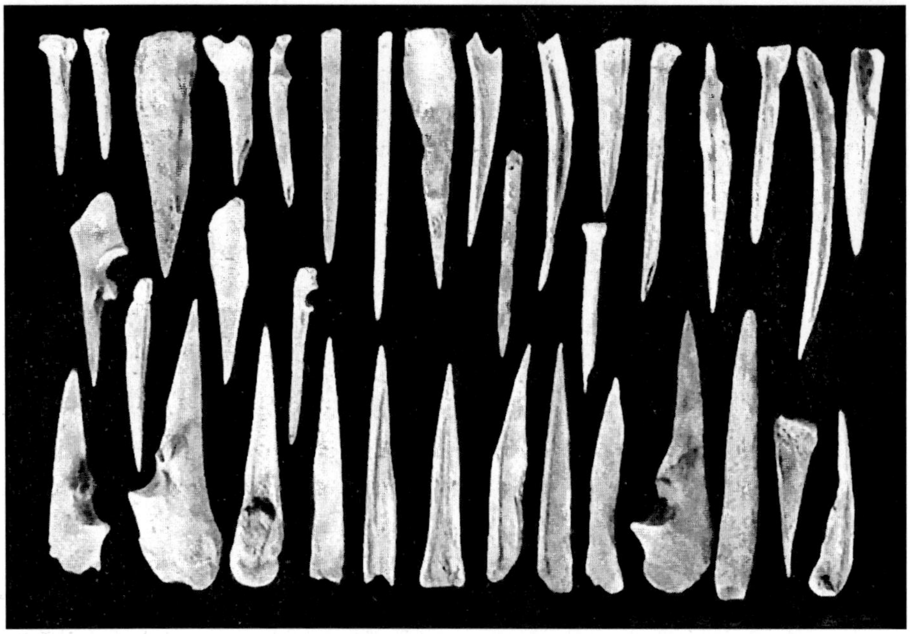

Figure 5.7. Photograph from Webb (1938) showing bone awls from Meredith Cave (plate 14a).

cord paddle marking.

15. Rims of vessels were plain and straight, without handles or lugs.

16. The entire absence of any evidence of the smoking custom.

From an inspection of the material evidence from this site one is driven to the conclusion that the occupants of this cave have no connection with the builders of the earth mound and village sites of the Basin, but that their cultural affinities are to be sought elsewhere.

However, besides the similarity in traits at this site with the rock shelter above referred to, some of the potsherds found in this cave are remarkably similar to those reported by Harrington from upper Tennessee River near Lenoir City, Tennessee, and classified as belonging to his "round-grave culture." He described this type of sherd as "marked with parallel corrugated indentations" and asserted a belief based on its association that it was not of Cherokee origin.

Claflin has found a similar pottery type at Stallings Island, Georgia, and similar sherds are reported by Bushnell from Kellys Ford, Virginia, on the Rappahannock. Bushnell expresses the belief that these "parallel corrugated indentations" as described by Harrington were produced by a basket used in forming the vessel. Both Harrington and Bushnell seem to regard this type of pottery as the oldest

5. The association with cremated remains of some and often many well-made artifacts. These offerings never include pottery.

6. The common use of two-holed flat and bar gorgets. Gorgets were made of hematite, cannel coal, steatite, slate, and limestone.

7. The finding of oval vessels of steatite, with end lugs, cut from a solid block.

8. The use of a wide range of stemmed and unstemmed projectile points. There was, however, an entire absence of small triangular points.

9. The considerable use of bone for awls and scrapers.

10. The use of bone fishhooks.

11. The use of cups and spoons cut from carapace of tortoise.

12. The finding of the femora of bear cut squarely off and hollowed out. These may have been used as handles, but some are found charred inside as if used as a torch containing grease.

13. Practically speaking, no use was made of worked shell.

14. Pottery was gravel tempered, of coarse texture, and usually had grass or

found in their respective regions.[2]

Unfortunately, Webb's study predates the use of Carbon-14 dating, so we do not have associated dates for this site. Spelean historian Angelo I. George noted that burials numbers 3, 5, 7, and 8 all have crushed skulls and that burials numbers 10, 12, and 13 are missing the skull. According to George, this could indicate that these people were sacrificed in some way.[3] A normal American Indian burial does not show a mutilated skeleton.

The story of the "burial wrapped in cane and bark resting on a ledge of rock in the cave wall" is particularly fascinating. This could be the description of an American Indian mummy. It is disappointing that it was removed from the cave sometime between its discovery in 1935 and Webb's visit in 1938. As a trained archaeologist, Webb could have given us a detailed description of the contents of this well-preserved burial and it would have been collected and preserved in the Smithsonian Institution. Unfortunately, we may never know what happened to this burial or what its contents were.[4]

Saltpeter Mining During the Civil War

Thomas C. Barr Jr's exploration of Meredith Cave indicates that it was worked extensively for nitrates during the Civil War. At the time of Barr's visit in the 1950s he reports, "The remnants of several wooden troughs and hoppers are still in the cave. Wagon tracks and markings where the walls were scratched by wagon wheels are still to be seen in the cave several hundred feet back of the entrance."

Webb (1938) also mentions that, "During the Civil War the Confederate Army used the cave as a source of saltpeter. Large piles of earth and ashes, the by-products of the process of extracting the niter from the soil, were carried out of the cave and dumped at the entrance. Excavation in the ash beds outside the cave yielded no evidence of prehistoric occupation."[5] He also goes on to mention the presence of "wooden vats used by the Confederates in leaching the soil for niter," but does not give their exact location or any further description of them.

It is certainly possible that Meredith Cave was also mined during the War of 1812. Tennessee, along with Kentucky, Virginia, and West Virginia, were mined extensively during this time period, but records of this mining activity are few and far between. Mammoth Cave in Kentucky and Big Bone Cave in Tennessee were two of the largest saltpeter mines in operation during the War of 1812.

A Description of the Cave

Meredith Cave is not open to the public at the present time. Barr's book, *Caves of Tennessee*, gives the following written description of the cave:

> Meredith Cave, also known as Saltpeter Cave, is one of the largest known caves in east Tennessee. It is a long, winding, and branching cave of the joint type. The entrance is a semicircular opening 40 feet wide and 12 feet high. Immediately back of the entrance is a spacious room about 190 feet long and from 40 to 60 feet wide. Back of this, most of the cave is much narrower. The combined length of all the penetrable passages is at least 5,149 feet. The average width for the entire cave is about 25 feet. The ceiling is unusually high, probably averaging between 60 and 70 feet. Most of the floor is dry, and in places it is dusty; in a few places, however, dripping water keeps it wet. Much of the floor is approximately level, but in several places it is necessary to

2 William S. Webb, *An Archaeological Survey of the Norris Basin In Eastern Tennessee*, Smithsonian Institution, Bureau of American Ethnology, Bulletin 118, 1938, 32 pages.

3 Angelo I. George, Personal communication, January 8, 2008.

4 For further information on American Indian mummies, read *Prehistoric Mummies from the Mammoth Cave Area* by Angelo I. George, Editor (1990). This book contains reprints of 35 early reports on the discovery of mummies in caves, including caves in Tennessee.

5 Ibid.

progress up or down steep slopes.

Several formations are present in Meredith Cave, and many are still actively developing. Most of the deposits are black in color, as a result of natural staining. During the Civil War the earth on the floor of the cave was worked extensively for nitrates. The remnants of several wooden troughs and hoppers are still in the cave. Wagon tracks and markings where the walls were scratched by wagon wheels are still to be seen in the cave several hundred feet back of the entrance. Numerous white chalk-like deposits occur in the drier portions of the cave.

In 1929 the cave was being used by the entire community as a storage place for fruits and vegetables. In the dry places, such perishable farm products as sweet potatoes are said to keep from year to year.

Several stairways were constructed in Meredith Cave about 30 years ago in an attempt to develop it as an attraction. These are now rotten and unsafe.

Archaeological investigations in Meredith Cave conducted by Webb (1938) revealed 13 burials near the entrance. Ash and midden debris were found to a distance of 600 feet from the mouth and some pottery and bone instruments were uncovered.[6]

Meredith Cave as a Commercial Cave

According to Barr (1961), "Several stairways were constructed in Meredith Cave about 30 years ago in an attempt to develop it as an attraction. These are now rotten and unsafe." It is quite possible that this commercial development was done about the same time that nearby Norris Lake was completed, attracting tourists to the region. Meredith Cave is currently owned by the same person who owns the nearby Shanghai Boat Dock, so perhaps the person who built the boat dock when the lake was created by Norris Dam attempted to develop Meredith Cave as a tourist attraction at the same time. We do not know what years the cave was open to the public, but Barr indicates that it may have been sometime during the 1930s. Currently, the cave is reported to be used for boat storage during the winter months by the owner of the nearby boat dock.

The Meredith Cave Beetle

Meredith Cave is home to the Meredith Cave Beetle (*Pseudanopthalmus sidus*). This tiny beetle (about one fourth of an inch long) is a true troglobite. Troglobites are animals that are so highly adapted to life in caves that they cannot live anywhere else. They spend their entire life in the totally dark cave environment. The Meredith Cave Beetle is known only from this one locality.[7]

Meredith Cave Today

Meredith Cave is a bat hibernaculum. Entrance is prohibited from September 1 through April 1 of each year. This allows the bats to hibernate undisturbed. Bats must live through the entire winter on their fat supply. This is possible by hibernating, since their body processes slow down and they are able to live on very little energy. If people enter such a cave and wake the bats from their hibernation several times during their hibernation period, the bat's metabolism will rise to its normal rate until they go back into hibernation. This can result in the bats starving to death before spring.

Directions

From Knoxville, go northwest on Interstate 75 for 34 miles to Exit 134. Go right (northeast) on U.S. Highway 25W for 6.3 miles. Turn right onto Hunters Branch Road. Go 0.5 mile, and then turn right onto Demory Road. Go 3.9 miles, and then turn right to stay on Demory Road. Continue 0.8 mile, then turn left onto Shanghai Road. After 1.3 miles you will reach the Shanghai

6 Thomas C. Barr Jr, *Caves of Tennessee*, Bulletin 64, Tennessee Division of Geology, 1961, pp 89–91.

7 Thomas C. Barr Jr, Personal communication, May 11, 2007.

Boat Dock. Approximately 0.2 miles before you reach the Shanghai Boat Dock, there is a private drive to the left that goes 0.3 miles to the entrance to Meredith Cave. However this is a private drive and Meredith Cave is on private property. Be certain that you have the necessary permission before driving down this private drive and before entering the cave.

Side Trips

On your return to Interstate Highway 75, you will pass the entrance to Cove Lake State Park half a mile before you reach the Interstate. This State Park borders on U.S. Highway 25W and the entrance is well marked. This park consists of 673 acres of land and is located on the eastern edge of the Cumberland Mountains. There is a 210-acre man-made lake, Cove Lake, and fishing is allowed. Flat bottom fishing boats are available for rent year round and pedal boats are available in the summer months.

Cove Lake State Park features a 106-site campground, a large picnic area, and a 3.6-mile paved hike and bike trail. In the winter months, the park is home to several hundred Canada geese.

Chapter 6

New Mammoth Cave

The Discovery of New Mammoth Cave

New Mammoth Cave is located approximately 35 miles west-southwest of Cumberland Gap. As was noted in Chapter 3, Daniel Boone followed the Warriors Path through the Cumberland Gap in 1775, paving the way for Caucasian settlers, along with their African slaves, to move west into Kentucky and Tennessee. It would be reasonable to assume that New Mammoth Cave was discovered shortly after settlers moved into this area. According to Berlen C. Moneymaker, the cave was mined for saltpeter during the War of 1812, so we know it was discovered before 1812. An inscription on the wall, dated May 26, 1819, records a party held at the cave, indicating that the cave was already a favorite recreational site with local residents.

New Mammoth Cave is located on the side of a major valley, Elk Valley, and the fertile valley bottom would have been quickly claimed and converted into farmland. Before being settled and converted into farmland, Elk Valley was a migratory path for the elk that lived in this area. These game trails became the transportation routes of the American Indians and the settlers. Many of the early saltpeter mines are positioned adjacent to these transportation routes.[1] The original name of the cave is unknown, but the name New Mammoth Cave was probably not coined until 1921 when the cave was opened to the public.

Considering the proximity of New Mammoth Cave to Cumberland Gap, it is interesting to note that the first person to make gunpowder in Kentucky was Daniel Boone. Professional geologist and spelean historian Angelo I. George provides the following information:

> On the Kentucky frontier, far inland from east coast markets, pioneer settlers, backwoodsmen, and explorers made do from what the land could provide. Two indispensable commodities needed on the western frontier were salt and gunpowder. Stockades and settlements sprang up along game trails near intersections with salt licks and springs. Caves and sandstone rock shelters were excavated and processed for saltpeter, first for personal use. By 1802, saltpeter and gunpowder manufacturing were fairly well established industries in Kentucky and neighboring states. The most significant sites were Mammoth Cave and Great Saltpetre Cave in Kentucky, Wyandotte Cave in Indiana, and Big Bone Cave in Tennessee. Collectively, the historic artifacts and documents associated with these caves present a coherent picture of life and working conditions in saltpeter mines.
>
> At first only a few people on the frontier knew the secrets of saltpeter and gunpowder manufacturing. Backwoods icon Daniel Boone and his contemporary, an African American, Monk Estill, are the first on record to have made gunpowder in Kentucky. As the region became more settled and industry developed, individuals such as John James DuFour; Fleming Gatewood Sr, Levi Brashear, Archibald Miller Sr, and African American

1 Angelo I. George, Personal Communication, November 13, 2007.

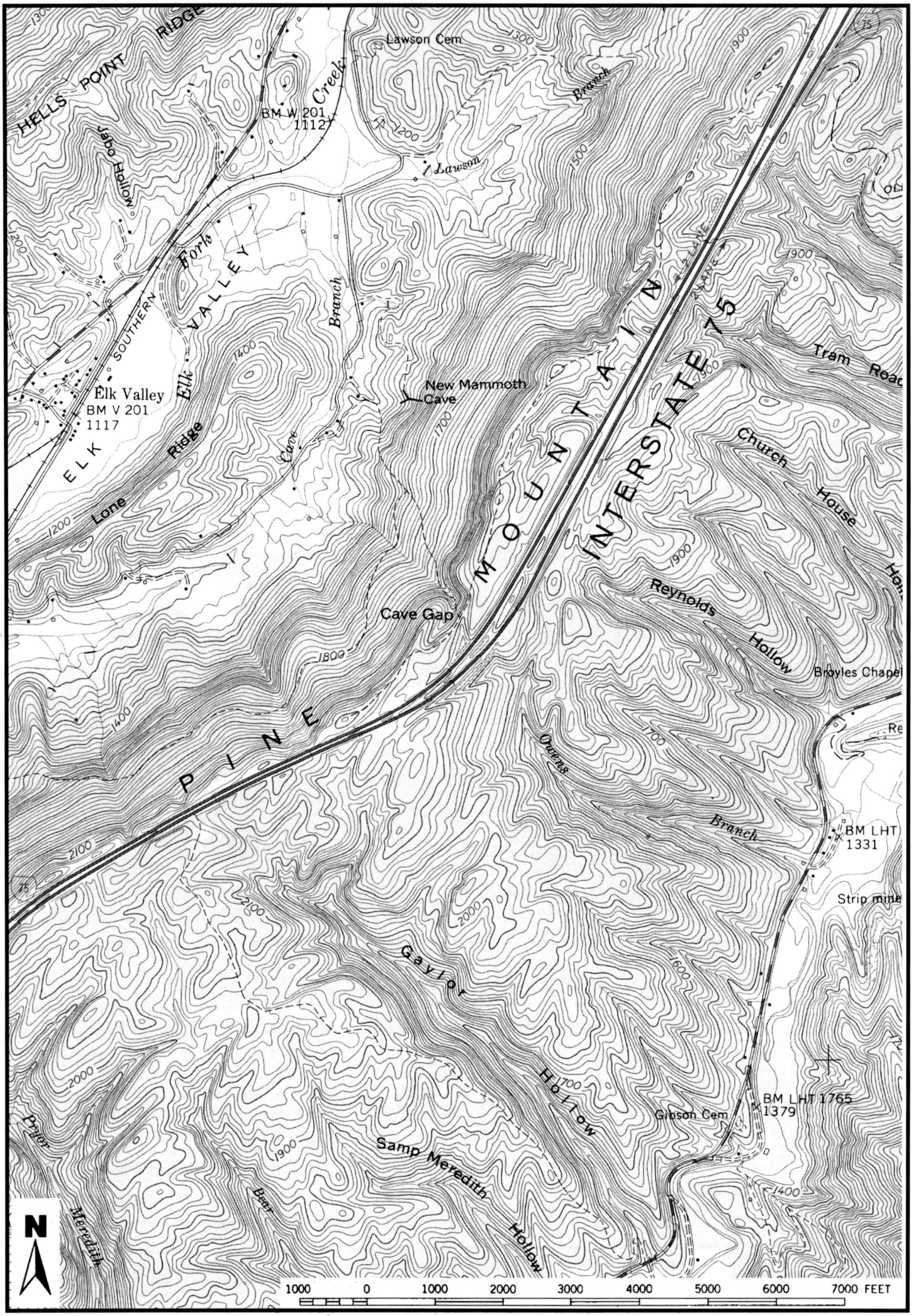

Figure 6.1. Topographic map showing the general location of New Mammoth Cave.

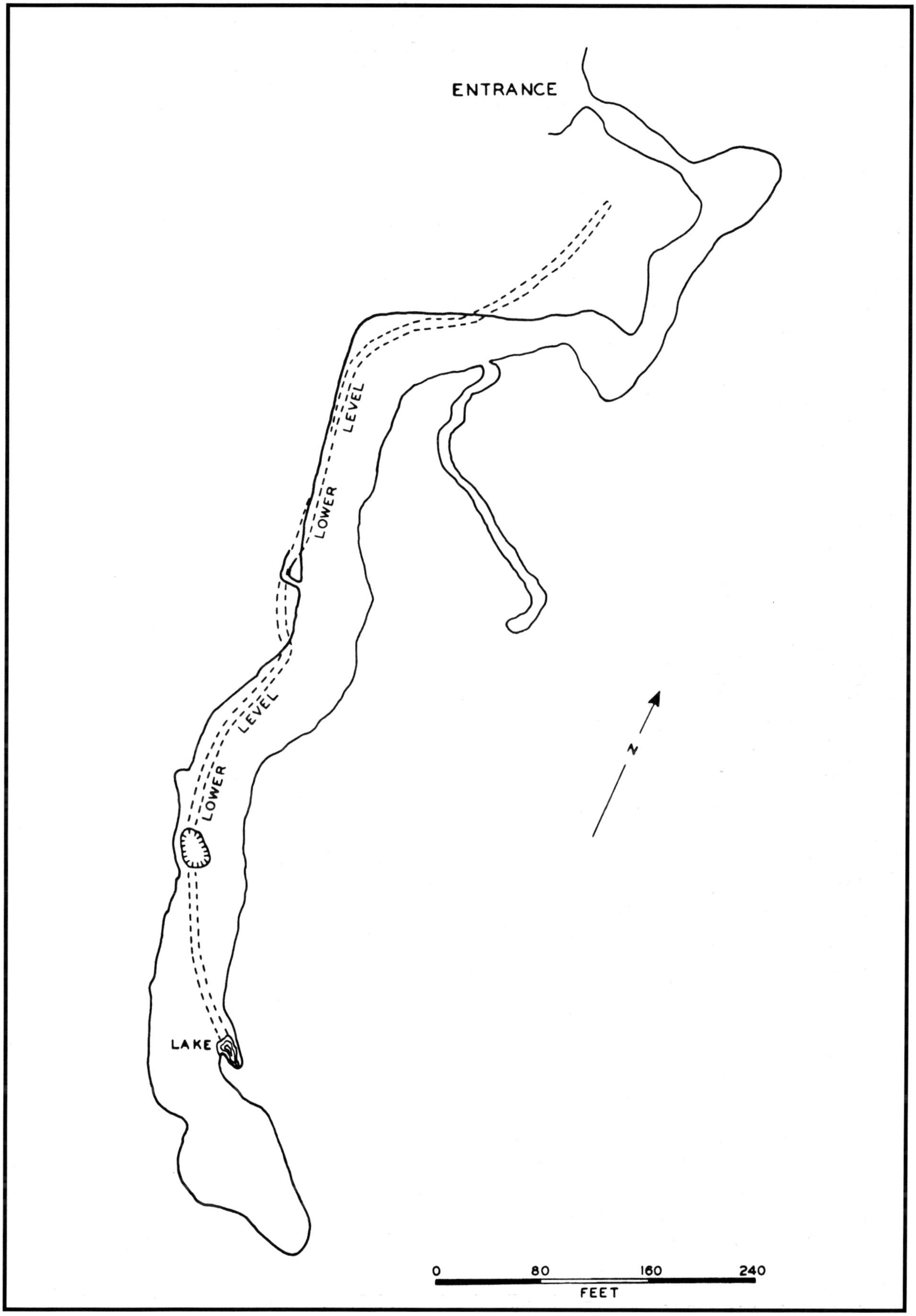

Figure 6.2. Map of New Mammoth Cave by Berlen C. Moneymaker, 1929, from Caves of Tennessee, *page 92.*

Figure 6.3. Bob Currie at the New Mammoth Cave Entrance in 1982 before the gate. Photo by Jim Whidby, August 28, 1982.

"Free" Frank McWorter made saltpeter in commercial quantities. At the height of the industry, there were hundreds of these "strolling chemists" as they were termed in 1809 by Lexington entrepreneur Samuel Brown, MD.[2]

George goes on to note that saltpeter mining began at Mammoth Cave in 1798, so saltpeter mining in New Mammoth Cave could also have begun by then. The War of 1812 created a large increase in the demand for gunpowder, but many of these mines had been in operation for up to 14 years before that conflict began.

Saltpeter Mining During the War of 1812

According to Moneymaker (1929), New Mammoth Cave was mined for saltpeter during the War of 1812. Cave historian Jim Whidby reports that saltpeter vats are still visible in the cave today, including both square-shaped and V-shaped vats.[3] Tom Barr, in the next section of this chapter, has reason to believe that New Mammoth Cave was also mined for saltpeter during the Civil War. The details of how saltpeter was mined from cave deposits are discussed in detail in Chapter 10 on pages 181 through 187.

The presence of both V-shaped and square-shaped vats indicates a significant mining

2 Angelo I. George, *Mammoth Cave Saltpeter Works*, H.M.I. Press, 2005, pp 2–3, 5.

3 Jim Whidby, Personal communication, August 1, 2007.

Figure 6.4. Mike Whidby examining remains of a V-shaped saltpeter vat inside New Mammoth Cave. Photo by Jim Whidby, February 10, 2007.

operation. Many mines began with the smaller, V-shaped vats, then added the larger square-shaped vats as production increased. The remains of these vats in New Mammoth Cave are poorly preserved. Much of the wood has rotted away, but the dirt that was inside these leaching vats remains as "casts," indicating their size and location.

Dead Air and Dried Rats

Caver Tom Barr wrote the following fascinating article about New Mammoth Cave in 1954. It describes both the 1929 visit by Berlen Moneymaker and Barr's own visit in 1954:

> In his thesis on east Tennessee caves, NSSer Berlen Moneymaker, Chief of Geology for TVA, mentioned finding dried rats in a side passage in Campbell County's New Mammoth Cave. New Mammoth is a fairly large cave (1,600 feet long, average 15 feet high, and 30 feet wide) in the Newman Limestone (Mississippian) on the east side of Elk Valley, 10 miles south of Jellico, Tennessee. The cave was formerly commercialized, in a half-hearted way with gasoline lanterns, and during the War of 1812 and the Civil War was extensively mined for saltpeter. Moneymaker attributed the mummification of the rats to lack of oxygen because of poor [air] circulation. This was of considerable interest, and had been cited by Ripley [*Believe it or Not*] and by the French speleologist, Norbert Casteret. Primarily for the purpose of seeing this passage, Tom Barr and Bert Denton journeyed to Oak Ridge, picked up Russ Hunt (groggy from drinking lemonade the night before), and after three or four false starts found themselves on the road to Elk Valley.

Figure 6.5. Richard Wallace and Mike Whidby by a mud cast of a square saltpeter vat. Photo by Jim Whidby, February 10, 2007.

Figure 6.6. Mike Whidby among mud casts of saltpeter vats. Photo by Jim Whidby, February 10, 2007.

Figure 6.7. Mike Whidby by stone steps inside New Mammoth Cave. Photo by Jim Whidby, February 10, 2007.

way to go by way of Jellico. New Mammoth and a small cave near Jellico appear to be the only caves in this part of Campbell County. The only prominent limestone outcroppings are found where the Newman has been exposed on the face of the Pine Mountain thrust fault, which forms the eastern side of Elk Valley.

The mouth of the cave is 6 feet wide and 10 feet high and is situated under an

The first road proved undesirable, because it was reported that it was possible to get down the mountain through the mud, but you couldn't get back up. Having slid toward many a precipitous embankment in the course of the mile-long stretch from the Oneida-Caryville Road to Pioneer, the three explorers conjured up visions of a speedy, uncontrolled descent of the mountain, and quite happily drove 60 miles out of the

Figure 6.8. Mike Whidby by another stone stairway inside New Mammoth Cave. Photo by Jim Whidby, February 10, 2007.

impressive rock shelter overlooking the valley floor 400 feet below. The cave trends S-30°-E for 1,130 feet and has 600 feet of lower level passageways and several stairways and bridges of wood. The rock dips toward the east at about 30 degrees, and the cave is full of breakdown. In places it is 20 feet high and 40 feet wide.

Figure 6.9. Sign painted on a wall inside New Mammoth Cave. "Visited Sunday, May 26, 1819... ." Photo by Jim Whidby, June 1983.

The side passage in which the dead air pocket had been reported was finally located. Its entrance is almost obscured by a wooden bench which had been constructed for the convenience of cavern visitors. Tom Barr explored the entire passage, which is a high dip tube 150 feet long, without experiencing any difficulty in breathing or noticing any change in the intensity of the carbide flame. No dead cave rat could be seen in the nest at the end, although a live one was seen on the wall. (It had been blinded in one eye a few days before by a kindly young fellow from the Baptist Mission's picnic to the cave.) Just as he was about to leave the passage, Tom noticed a dried carcass of a cave rat under the steps leading down into the tube. It was collected for future examination. Study of the carcass appears to show that it does not differ in any way from dried animals that one often finds in xeric cavern passages. It was concluded that microenvironmental conditions in this portion of the cave are not suitable for the growth and development of decay-causing bacteria.

Thus there is at present no low oxygen pocket in New Mammoth Cave, and present evidence may be readily explained without recourse to the idea that the oxygen content of the passage was low in the past. The deep layer of debris in the cave rat's nest at the end of the passage indicates that the air circulation has been quite good for some time. We hate to refute Mr Moneymaker and contradict Mr Ripley, and N. Casteret, but facts *is* facts! A bilious old guide in old Mammoth Cave once warned us to be sure and take along a gasoline lantern in holes south of the border (Kentucky, that is) because, "There's bad air in some of them Tennessee caves." We're still looking![4]

Figure 6.10. Richard Wallace at the old-style gate inside the entrance to New Mammoth Cave. Photo by Jim Whidby, June 1983.

It is interesting to note that there was considerable evidence of commercial development in the cave at the time of Barr's visit in 1954, including stairways, bridges, benches, and steps.

Moneymaker read Tom Barr's article in the *NSS News* and wrote the following response:

> The rat remains and the oxygen-deficient air were both found at the terminus of a narrow, crooked cul-de-sac which was not noticeably low in oxygen except in the back 50 feet and in the terminal chamber. Oxygen content was so low that breathing was difficult, candle flames were extinguished and lighted matches went out immediately. The writer never stated nor implied that this was a

4 Thomas C. Barr Jr, "New Mammoth's Dead Air & Dried Rats," *NSS News*, vol 13 (1955), no. 2, p 4.

> natural condition.
>
> Although Mr Barr stated that "air circulation has been quite good for some time" at the time of the writer's survey, the air beyond the last bend in the passage was, for all practical purposes, both static and stagnant—which is not unusual in a dead end passage.[5]

Although oxygen-deficient air is rare in Tennessee caves, it is not impossible. Both writers describe a downward-sloping, dead-end passage. Since carbon dioxide is heavier than air, it is certainly possible that carbon dioxide, or even carbon monoxide, may have settled into this passage if open flames were being used in the cave for light, especially over an extended period of time.

A 1983 Visit to the Cave

Cave historian Marion O. Smith visited New Mammoth Cave on June 1, 1983. He was accompanied by cavers Richard Wallace and Jim Whidby. Below is the fascinating account of his visit to the cave:

> We actually entered the cave at 8:09 P.M. (EDT). The entrance is about 10 feet high and 8 feet wide in a bluff with a stream resurging below. Including the lower level, there is about 1,800 feet, plus, of passage present. The upper level is dry, generally quite wide, up to 50 feet wide and up to 20 feet high. This cave was *heavily mined for saltpeter*, allegedly during the War of 1812, although we saw no dates that early. It is virtually impossible to count the vat sites since they run into each other. I got 24 but that is probably a low estimate. The piles of dirt generally were quite large. In a couple [of] places near the entrance a lot of used dirt (mud) had been thrown on shelves. There were a good many piled rocks, one pile was stacked and 3½ feet high. There were soot marks, scattered tally marks, and in three places (at least) rock steps. In the lower stream level rocks were also piled and there were rock "bridges" across the creek. Near the end of the cave, at the stream level, was a concrete dam—20th century—perhaps built during or after 1921 when the cave was opened to the public.
>
> Other than "1862" with an illegible name, I found no wartime names or dates. Others were "Samuel D____ 1868," "A.P. Dalton Bornd _____ Ohio 1860," "John Thies" [probably recent], "Demoros Key 1886," "F.M. Baird was here and (?) 1880." I attempted a number of photos—mainly with Keith Hausman in them. We got out of the cave about 10:14 P.M.[6]

Clearly, New Mammoth Cave was a large, industrial saltpeter mining operation. Only ten other Tennessee caves are known to have had this large-scale type of operation.[7] Unfortunately, it is still unknown if this cave was mined during the War of 1812, the Civil War, or both.

A Description of the Cave

New Mammoth Cave is not open to the public at the present time. The book, *Caves of Tennessee*, which was written by Thomas C. Barr Jr in 1961, gives the following written description of the cave:

> New Mammoth Cave is a large strike gallery 1,100 feet in length. The mouth is 5½ feet in diameter and overlooks the valley. The floor of the cave is covered with huge slabs of breakdown, at the bottom of which is a small stream. About 600 feet of "passage" may be traversed under the breakdown. The average distance from the top of the breakdown to the arched roof of the cave is 15 feet, and the average width

5 Berlen C. Moneymaker, "Abstract of Letter to the Editor From Berlen C. Moneymaker," *NSS News*, vol 13 (1955), no. 5, p 6.

6 Marion O. Smith, Personal Communication, November 1, 2007.

7 For further information on saltpeter mining in Tennessee, see *Big Bone Cave* by Larry E. Matthews (2006). Chapter 28 is titled: "Saltpeter Mining in Other Tennessee Caves."

Figure 6.11. Cluster of bats hibernating inside New Mammoth Cave. Photo by Jim Whidby, February 10, 2007.

of the main passage is 30 feet. One side passage is a narrow opening which extends down dip for 150 feet. Few formations are developed, although the cave is damp in several places.

According to Moneymaker (1929) the cave was worked for niter during the War of 1812. In 1921 it was opened to the public by John Brown as "New Mammoth Cave." The name was changed to "Cumberland Mammoth Cave" in 1928, when H.D. Blankenship took over the operation of the cave. It is no longer open to tourists.

Moneymaker (1929, 1930) recorded an apparent low concentration of oxygen in the side passage. The anaerobic condition was temporary, however, because it did not exist when the writer visited the passage in 1954, but in 1928 there was difficulty in breathing (Moneymaker, personal communication). Mummified cave rats (*Neotoma magister* Baird) were found in this passage, the low humidity probably having inhibited bacterial decomposition.[8]

According to Tom Barr, this cave was shown under two commercial names: **New Mammoth Cave** and **Cumberland Mammoth Cave**. The cave opened to the public in 1921 and closed sometime after 1928. The stock market crash of 1929 and the Great Depression that followed put a number of commercial caves out of business. People no longer had the money to travel to these sites.

Today, it is hard to imagine that anyone would have attempted to open a commercial cave in this remote and rugged area. However, a main railroad line runs through Elk Valley and the entrance to New Mammoth Cave is only a little over a mile from where the train once stopped in the community of Elk Valley. Perhaps the developer, John Brown, hoped that enough people would stop and visit his cave to make it a commercial success. It would appear

8 Thomas C. Barr Jr, *Caves of Tennessee*, Bulletin 64, Tennessee Division of Geology, 1961, pp 91–93.

Figure 6.12. Mike Whidby (L) and Jim Whidby (R) by the new, bat-friendly gate at the entrance to New Mammoth Cave. Photo by Richard Wallace, February 10, 2007.

that the cave was open for the entire decade of the 1920s, but it is uncertain in what year it closed.

The Bats of New Mammoth Cave

New Mammoth Cave is home to five species of bats during the winter hibernation season. This includes the federally endangered Indiana bat. Hibernating bats are very sensitive to disturbance during their hibernation, since they must live through the winter on stored fat. If persons enter a cave several time during this hibernation season and wake the bats up, this speeds up their metabolism until they go back into hibernation. Even being disturbed a few times can use up enough of this fat supply to cause a bat to starve to death before the end of winter. If you enter a cave during the winter and observe hibernating bats, please leave immediately.

In order to protect the bats that depend on this cave for a hibernation site, a gate was built on the entrance in 1980. Thanks to the protection that the Indiana bat is receiving at this and several other caves, it now appears to be coming back from the brink of extinction.

New Mammoth Cave Today

Currently, New Mammoth Cave is located on private property and is not open to the public. Despite the fact that it is located only 35 miles from downtown Knoxville and within half a mile of Interstate Highway 75, it is located in a rural mountain valley (Elk Valley) near the foot of Pine Mountain and it is not close to an Interstate Highway exit.

Since New Mammoth Cave is an important habitat for bats, it is gated and the U.S. Fish and Wildlife Service controls access. As a bat sanctuary, this cave should be considered closed to the general public. Please do not attempt to visit this cave without permission from the

Figure 6.13. Political statement on a wall of New Mammoth Cave. Photo by Jim Whidby, June 1983.

federal authorities.

Directions

From Knoxville, go northwest on I-75 for 41 miles to the Oneida/Huntsville Exit (Exit 141). Turn left onto Tennessee Highway 63 and proceed 4.5 miles to the intersection of Tennessee Highway 297. Follow Tennessee Highway 297 for 8.0 miles to the community of Elk Valley. Turn right onto Elk Main Street. This road affords an excellent view of an impressive ridge known as Pine Mountain, located approximately one mile to the east. If you follow this road, it loops south back to Tennessee Highway 297. Pine Mountain rises 1,000 feet above the valley floor. New Mammoth Cave is located 220 feet uphill on the side of Pine Mountain. More specific directions will not be given, since this cave is closed to the public.

Side Trips

Both Highway 297 and Interstate Highway 75 follow Elk Valley and Pine Mountain through a very rugged area of Tennessee. In prehistoric times, migrating animals (hence the name, Elk Valley) and American Indians used this same route. In Tennessee, much of this area has been acquired by the state for use as a Wildlife Management Area and elk, which have been absent in Tennessee for nearly 200 years, have recently been introduced. Hopefully, in a few years the elk herd will expand and spread out across its former range in these mountains.

If you continue north on Highway 297 for an additional 9 miles you will come to the town of Jellico, which sits on the Kentucky–Tennessee border. This is classic, Appalachian coal mining country and you will see numerous scars on the mountainsides from past and current strip mining. However, the areas that have not been mined are very beautiful and well worth seeing.

At Jellico, you can turn right (east) onto Highway 9, which will lead you back to Exit 160 of Interstate Highway 75 after about 2 miles. Turn right (south) onto Interstate 75 and proceed south for 60 miles to Knoxville.

Chapter 7

Alum Cave

The Discovery of Alum Cave

The recorded history of Alum Cave goes back to 1837 when three farmers (Ephraim Mingus, Robert Collins, and George W. Hayes) from Oconaluftee, North Carolina, applied at the Sevier County Land Office for a grant of a 50-acre tract of land that would include Alum Cave and its salt deposits. Tennessee sold the tract of land to the men on December 6, 1838, and the Epsom Salts Manufacturing Company was subsequently formed to mine the deposit. The minerals that were recovered in commercial quantities were alum, Epsom salt, saltpeter, magnesia, and copperas. According to the old records, many tons of alum and Epsom salt were recovered. A leaching system consisting of water vats and log hoppers was utilized to separate the salts from the deposits covering the floor of the rock shelter. The easily accessible salts were depleted by the mid-1840s. The company was sold and mining activity decreased until the beginning of the Civil War, when demand for these salts (and for saltpeter) again made the site profitable to mine. No further mining is known to have occurred after the Civil War. [1]

The earliest written account of Alum Cave was made by Charles Lanman (1819–1895) in an article titled "Letters from the Allegheny Mountains" that was published in the September 1850 issue of the *Southern Quarterly Review*. The date of this visit was about 1848. The following account of that visit to Alum Cave was published in a book by Lanman, in 1856, entitled *Adventures in the Wilds of the United States and British American Provinces*:

> The ordinary time required to visit the Alum Cave is two days: but, owing to bad weather, my friend and myself occupied the most of four days in performing the same trip. To give a minute account of all that we met with would occupy too much time, and I will therefore only record in this place the incident which made the deepest impression on my own mind.
>
> Our first night from home we spent in the cabin of a man who treated us with the utmost kindness, and would not receive a penny for his pains. So much for mountain hospitality. And now, to prove that our friend was an intelligent man, it may be mentioned that he is adept in the following professions and trades, viz. those of medicine, the law, the blacksmith, the carpenter, the hunter, the shoemaker, the watchmaker, the farmer, and he also seemed to possess an inkling of some half dozen sciences. Now, I do not exactly mean to assert that the gentleman is a master practitioner in all these departments of human learning and industry; but if you were to judge of his ability by his use of technical words, you would not for a moment imagine he could have a competitor. But so it is in this wild region, one man has to perform the intellectual labor of a whole district; and what is really a hard case, the knowledge which is thus brought to so good market is nearly always the fruit of a chance education and not a systematic one...

1 T. Dennis Coskren and Robert J. Lauf, "The Minerals of Alum Cave Bluff: Great Smoky Mountains, Tennessee," *Mineralological Record*, March/April 2000.

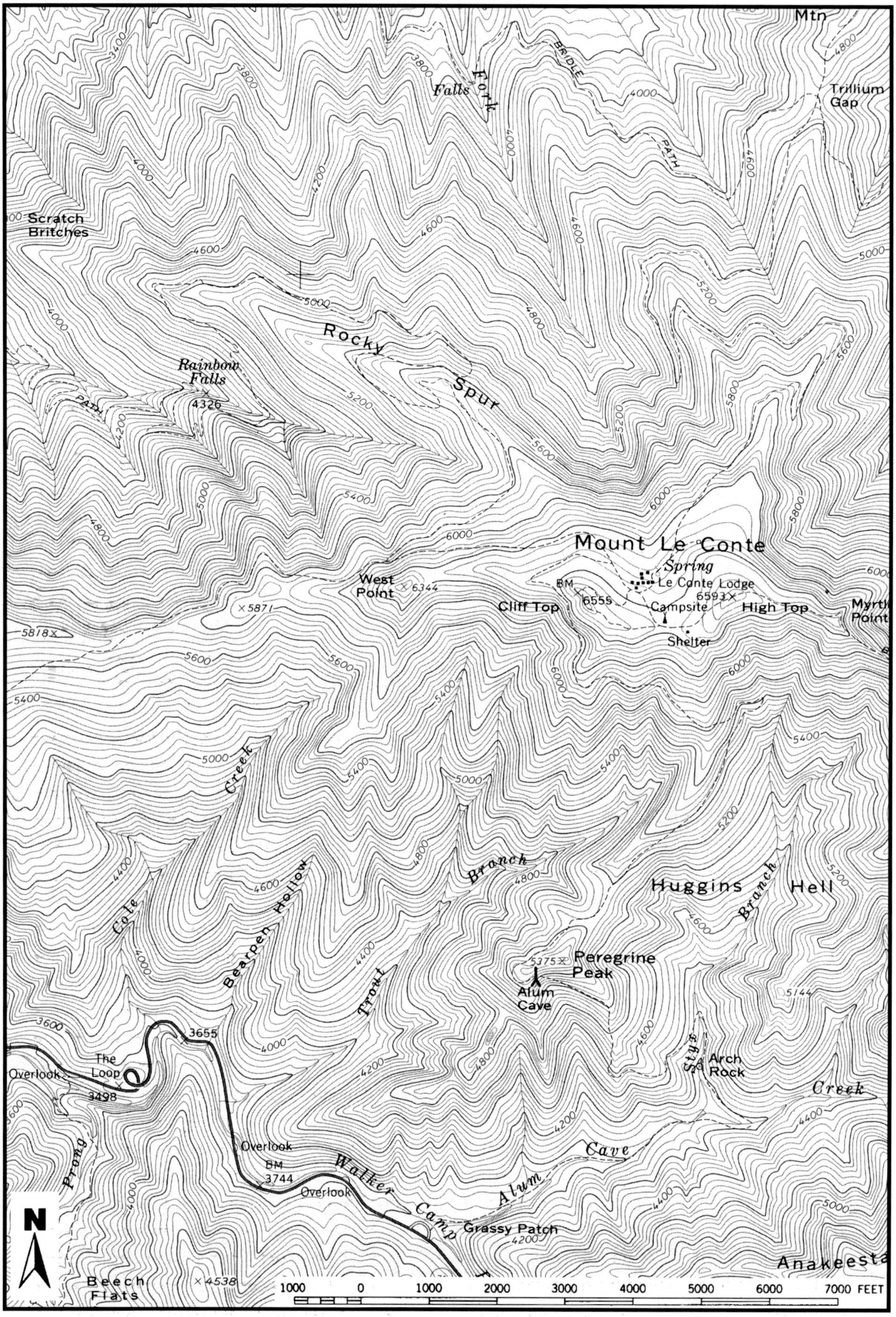

Figure 7.1. Topographic map showing the location of Alum Cave.

Figure 7.2. Illustration of "Alum Cave, Tennessee" from page 328 of the October 30, 1869, issue of Appletons' Journal: A Magazine of General Literature. *Courtesy of Donald B. Ball.*

On the following morning we traveled to the foot of Smoky Mountain, and having obtained a guide, who happened to be one of the proprietors of Alum Cave, we resumed our journey. In the immediate vicinity of the cave we came across an Indian camp, where were two Indians who were out bear-hunting. We were admitted under their bark roof, and with them spent the night, sleeping upon the ground. We remained a sufficient length of time to enjoy one supper and one breakfast...

On questioning our Indian landlord, as we sat around our watch fire, with regard to the Alum Cave, I could only gather the fact that it was originally discovered by the famous Chief Yo-na-gus-ka, who happened in his youth to track a bear to one of its corners, where he had a den. Disappointed on this score, I then turned to our guide to see what he could tell me about the cave that was not connected with its minerals, and the substance of his narrative was as follows:

"I hav'n't much to say about the cave that I knows of excepting one or two little circumstances about myself and another man. The first time I come here it was with my brother and two Indians. The sight of this strange gash in the mountain and the beautiful scenery all around made me very excited, and I was for climbing on top, and no mistake. The Indians and my brother started with me up the ledge at the north end of the cave, but when we got up half way, just opposite to an eagle's nest, where the creatures were screaming at a fearful rate, then all three of them backed down, and said I must not keep on. I told 'em I was determined to see the top, and I would. I did get on top, and after looking round a while and laughing at the fellows below, I began to think of going down again. And then it was that I felt a good deal skeered [sic.]. I found I couldn't get down the way I got up, so I turned about for a new place. It was near sundown, and I hadn't yet found a place that suited me, and I was afraid I'd have to sleep out alone and without any fire. And the only way I ever got down was to find a pine tree that stood pretty close to a low part of the ledge, some three hundred yards from the cave, when I got into its top, and so came down among my friends, who said it was a wonder I hadn't been killed.

I generally have had to pilot all strangers to the cave since that time, and I remember one circumstance that happened to a Tennessee lawyer, who caused us a good deal of fun; for there was a party of young gentlemen there at the time. We had a camp right under the cave, where it's always dry, and about midnight the lawyer I mentioned suddenly jumped up as we were all asleep, and began to yell in the most awful manner, as if something dreadful had happened. He jumped about as if in the greatest agony, and called on God to have mercy on him, for he knew he would die. O, he did carry on at a most awful rate, and we thought he must have been bitten by some snake, or was crazy, so we tore off his clothes to see what was the matter; and what do you suppose we found? Nothing but a harmless little lizard, that had run up the poor man's legs, all the way up to his arm-pits, thinking, I suppose, that his clothes was the bark of a dead tree. After the trouble was all over, the way we laughed at the fellow was curious.

Our second day at the Alum Cave (and third one from home) was a remarkably cheerless one; for a regular snow-storm set in, mingled with hail, and, before we could reach our horses and descend the Smoky Mountain, some three or four inches of snow had fallen. We spent that night under the roof of our good friend and worthy man, the guide, and it was with difficulty that we could induce him to receive a bit of money for all his trouble in piloting us and treating us to his best fare. On that night we ate our supper at nine o'clock, and what rendered it somewhat peculiar, was the fact that his two eldest daughters, and very pretty girls withal, waited upon us at table, holding above our heads a couple of torches made of the fat pines. That was the first time that I was ever waited upon in so regal a style, and more than once during the feast did I long to retire in a corner of the smoky and dingy cabin to take a sketch of the romantic scene. At sunrise on the following morning my companion and myself remounted our horses, and in three hours were eating our breakfast in Qualla Town.

In a latter article, titled "Novelties of Southern Scenery. III" and published in the October 1869 issue of *Appleton's Journal*, Lanman gives the following description of Alum Cave:

The mountain here mentioned is one of the most imposing in the Alleghany range, and its foundations lie both in North Carolina and Tennessee; and its chief attraction is a singular cliff known throughout that region as the Alum Cave. In reaching this spot, which is on the Tennessee side, you have to leave your horses near the top of the mountain, and then journey on foot for six miles up and down, far up and far down, over every thing in the way of rocks and ruined vegetation which Nature could devise, until you come to a mountain side which is only two miles, in a direct line, from your starting-place. Roaring along at the base of this mountain is a small stream, from which you have to climb a precipice in a zigzag way, which is at least two thousand feet high, when you find yourself on a level spot of pulverized stone, with a rocky roof extending over your head a distance of perhaps sixty feet. The length of this hollow in the mountain, or "cave," as it is called, is nearly four hundred feet, and, from the brow of the beetling precipice to the level below, the distance is about one hundred and fifty feet. The top of the cliff is covered with a variety of rare and curious plants, and directly over its centre trickles a little stream, which forms a pool, like a fountain, in front of a spacious piazza. The ingredients of the rock composing this cliff are alum, Epsom salt, salt petre, magnesia, and copperas, and the water which oozes therefrom is distinguished for its strong medicinal qualities. This strange and almost inaccessible, but unquestionably very valuable, cave belongs to an organized company, and, before the late war, had been worked with considerable profit, on account of its alum. The scenery upon which this cave looks down is also decidedly novel and interesting. From one point of view the mountains descend abruptly from either side, into a kind of amphitheatre, where the one on the right terminates in a very narrow and ragged ridge, which is without vegetation, while far beyond, directly in front of the cave, rises a lofty and pointed mountain backed by some three or four peaks of inferior magnitude. The ridge alluded to is very high, but yet the cave looks down upon it, and it is so fantastic in its appearance that, from different points of view, may be discovered natural holes, or windows, opening through the entire wall, while from other points of view the great rocky mass resembles a ruined castle, a decayed battlement, or the shattered tower of a huge cathedral. To gaze upon this prospect at the sunset hour, when the mountains are tinged with a rosy hue, and the great hollow, or basin, before you is filled with a purple atmosphere, and the rocky ledge is basking in the sunlight, like a huge monster on the bosom of a placid lake, affords one of the most curious and impressive scenes imaginable. But the locality, under any of its phases, will amply repay the lover of fine scenery for a long pilgrimage. When the writer visited this spot, which was in the month of May, the weather in the valley where he was staying was balmy and summer-like, but after leaving the cave on his return he had to wade through snow several inches deep... .

Figure 7.3. Postcard showing Alum Cave. From the Speleo Archives of Gary K. Soule.

Donald B. Ball reports that geologist James M. Safford (1822–1907) visited Alum Cave in 1855. He describes the cave as a source of alum and Epsom salt. Ball also located the following notice in the June 17, 1862, issue of the *Weekly Columbus* [Georgia] *Enquirer*:

Epsom Salts.—Messrs. Sensabaugh, Mingus and Long sent us a specimen of Epsom Salts manufactured by them

> from a cave on Smokey Mountain, [*sic*] between N. Carolina and Tennessee. They are now making 300 lbs. of Epsom Salts, and 400 lbs. of Alum daily. The salts are said to be superior to any heretofore sold in the South, and the Alum is equal. The manufacturers say they will be able to supply the whole Southern Confederacy with these necessary articles. Any one interested can take the Salts sent us, and try their effects.—Augusta Chronicle.

Another prominent early visitor to Alum Cave was Arnold H. Guyot, professor of geology and geography at Princeton University. The following article was located in the Great Smoky Mountains National Park Library, at Sugarlands Visitor Center:

> In 1858 Arnold H. Guyot, professor of geology and geography at Princeton, surveyed Alum Cave at 4,971 ft. MSL. Safford, whose report was not published until 1869, noted a wagon load each of Epsom salts and alum. He collected fine specimens "white and pure, a cubic foot in volume." Safford was impressed by the natural formation of these compounds, especially their purity and the inaccessibility of this "rockhouse."
>
> This area formed about 4,000 to 570 million years ago. In an orogenic, low grade metamorphic environment, shale (clay minerals) became slate with pyrite inclusions. In the weathering process ground water seeps through the slate and transports minerals to the surface of the cave where contact with oxygen (air) causes a chemical reaction creating these compounds. Alum, not a specific substance, refers to Double Sulfate Salts.
>
> Epsom salt is magnesium sulfate.

Figure 7.4. Mount LeConte as viewed from Highway 441. Photo by Larry E. Matthews, October 21, 2007.

Figure 7.5. Informational sign at the Alum Cave Bluffs Trailhead. Photo by Larry E. Matthews, October 21, 2007.

Magnesium, our second lightest metal and the eighth most abundant element in the earth's crust, is very active; thus it is never found free in nature. Although it was not isolated until 1808 by Davy, Grew extracted it from mineral water in 1695 in Epsom, England, hence the name Epsom salts.

Saltpeter is potassium nitrate, KNO_3. Known in the West since about the twelfth century, it is a major ingredient in gunpowder. During the Civil War the Epsom Salts Mfg. Co. revived to fill the military needs of the Confederacy. In an 1863 manuscript Professor Guyot advised the federal government to seize the Southern Appalachians, which he concluded were of vital importance to the Confederates. Indeed, they needed the saltpeter for gunpowder and the minerals for hospitals to replace the standard medicines which were contraband. Later in the war mining became impossible. Mining operations never reopened after the war, and in 1934 Alum Cave became part of the Great Smoky Mountains National Park. Now it is a favorite hiking destination on the shortest trail up Mt. LeConte.[2]

Donald B. Ball sums up the early history of Alum Cave as follows:

> Despite this temporary reprieve from commercial oblivion, no evidence was encountered to suggest that this formation attracted any significant amount of attention from either visitors or entrepreneurs following the Civil War

---

2 The Wee Roving Reporter, *The Cave That Isn't a Cave*, Source unknown. Copy on file at the Great Smoky Mountains National Park Library.

save for likely a very few adventuresome souls exploring the mountains. Alum Cave would attract little notice until some time after the establishment of the Smoky Mountains National Park.[3]

A Description of the Cave

Alum Cave is not a true cave. It is what geologists refer to as a *Rock Shelter*. A rock shelter is a rock overhang that resembles a cave entrance, but does not open into an actual cave. Alum Cave is an exceptionally large rock shelter and it is a favorite hiking objective of many visitors to the Great Smoky Mountains National Park. An excellent trail leads from the main park road, U.S. Highway 441, to Alum Cave and continues on to the summit of Mount Le Conte.[4]

Despite the fact that Alum Cave is not a true cave, at least as defined by cave scientists, it is included in this book due to its name, its popularity as a hiking destination in the Great Smoky Mountains National Park, and its historical interest.

Alum Cave Today

Alum Cave is located on one of the most popular trails in the Great Smoky Mountains National Park. The Alum Cave Trail begins at a parking lot along U.S. Highway 441, which is also known as the Newfound Gap Road, in the Park. This is the main highway that connects Gatlinburg, Tennessee, to Cherokee, North Carolina. Even though there is a large parking lot, it is frequently full.

The Alum Cave Trail goes 2.2 miles to Alum Cave, then continue another 3.3 miles to the top of Mt. LeConte. This is the shortest trail up Mt. LeConte (5.5 miles long), which accounts for part of its popularity. The elevation gain from the parking lot to Alum Cave is 1,120 feet. The parking lot is located at an elevation of 3,830 feet above sea level and Alum Cave is located at

Figure 7.6. The Alum Cave Bluffs Self-Guiding Nature Trail Brochure for 2007.

an elevation of 4,950 feet above sea level.

This is a moderately difficult trail, so be sure that you are physically fit enough for

3 Donald B. Ball, *Historical Notes on Alum Cave, Sevier County, Tennessee*, 11 pages.

4 Larry E. Matthews, *Descriptions of Tennessee Caves*, Bulletin 69, Tennessee Division of Geology, 1971, pp 82.

the 4.4 mile round trip hike and the 1,120 foot elevation gain. Weather can change suddenly in the mountains, so be prepared for inclement weather, dress appropriately, and take adequate water and snacks.

Figure 7.7. Alum Cave Bluffs about 1935. From an old postcard.

Directions

From Gatlinburg, Tennessee drive south on U.S. Highway 441. You will definitely want to stop at the Sugarlands Visitor Center, which is on the right shortly after you enter the Park. Here you can obtain information on the Park and view interpretive exhibits. There is an excellent selection of books and maps on sale here. There are also restrooms and water fountains. Proceed 8.8 miles on U.S. Highway 441 past the Visitor Center to the Alum Cave Trail parking lot. Follow the well-marked trail for 2.2 miles to Alum Cave. At the trailhead you can purchase a brochure titled: "Alum Cave Bluffs Self-Guiding Nature Trail." This brochure is published by the Great Smoky Mountain Association in cooperation with the National Park Service.

Side Trips

The Great Smoky Mountains National Park is the second largest national park east of the Mississippi River. It is the most visited park in the entire national park system. Located in a temperate climate, the Great Smoky Mountains National Park is a wonderful destination at any time of year. Don't let the fact that this is a heavily visited park discourage you. Traffic on the roads will be heavy at some times of the year, especially during the fall colors, but if you will park your car and follow any of the hundreds of miles of trails in the Park, you will soon escape the crowds. Most of the Great

Smoky Mountains National Park's visitors drive through, staring out of their car windows, and never venture more than a few hundred feet from the road. Park your car, start hiking, and enjoy this magnificent park.

The Alum Cave Trail continues another 3.3 miles past Alum Cave to the top of Mount LeConte. There is a rustic lodge located on top of Mount LeConte, the Mount LeConte Lodge, which has been in operation since 1926. It was built before the National Park was established in 1934. If you wish to stay at the Lodge overnight, reservations are required and you will need to apply well in advance. However, for the hardy hiker, it is possible to hike to the top of Mount LeConte and back down in one day. If you spend the night on top of Mount LeConte, you may wish to view the sunrise from Myrtle Point. This view looks east into the heart of the park. Another outstanding overlook is known as Clifftop. This view looks southwest, down the Oconaluftee Valley, and is a great place to watch the sunset. A shelter is available for backpackers, but reservations are required.

Figure 7.8. Alum Cave Bluff patch. From the Speleo Archives of Gary K. Soule.

Figure 7.9. Postcard showing Alum Cave. From the Speleo Archives of Gary K. Soule.

Chapter 8

Forbidden Caverns

The Discovery of Forbidden Caverns

The commercial entrance to Forbidden Caverns is in a small, abandoned quarry and the visitor center has been built over this entrance. Tourists enter the cave from inside this building. However, at one time there was a natural entrance to this cave further up the hillside. There is evidence that the cave was explored by Eastern Woodland Indians hundreds of years ago.

The cave is in pristine condition, without names written on the walls and ceiling and little other evidence of past exploration. This may be an indication that the natural entrance was small and uninviting to the early settlers. The earliest record of modern exploration is in 1919. As a result, when quarrying opened the cave in the early 1960s, it had the unmarred beauty and size to be an ideal cave to open to the public. Due to its proximity to the Great Smoky Mountains National Park, the cave is a popular tourist destination.

A Description of the Cave

Forbidden Caverns is currently open to the public and is one of the premier tourist attractions in the vicinity of the Great Smoky Mountains National Park. The book, *Descriptions of Tennessee Caves*, which was written by Larry E. Matthews in 1971, gives the following written description of the cave:

> Forbidden Caverns is a rather attractive and interesting commercial cave. The commercial portion consists of approximately 1,400 feet of passages. From the entrance the cave trends west for 80 feet and averages 10 feet wide and 7 feet high. At this point the largest room of the cave is encountered. This room is 150 feet long, as much as 70 feet wide, and as high as 20 feet. Two passages leave this room, and both trend roughly east-southeast. The northern passage is 500 feet long and averages 10 feet wide and 8 feet high. The southern passage is 450 feet long; the first half of this passage averages 10 feet wide and 8 feet high, and the last half averages 5 feet wide and 25 feet high. These two passages join and continue with the dimensions of the southern passage for 225 feet. This passage averages 10 feet wide and 30 feet high.
>
> Several of the passages leading from the commercial section are reported to be fairly extensive.
>
> Forbidden Caverns was commercialized soon after its discovery, and for this reason it is free of vandalism. There are many attractive formations throughout the cave.
>
> Forbidden Caverns is believed to be the downstream end of nearby Cuyler Cave, although it may prove impossible to connect the two.[1]

In 1964, a group of local businessmen began the development of the cave so that it could be shown to the public. This took three years to accomplish and the cave opened to the public in June 1967.

---

1 Larry E. Matthews, *Descriptions of Tennessee Caves*, Tennessee Division of Geology, 1971, pp 83–86.

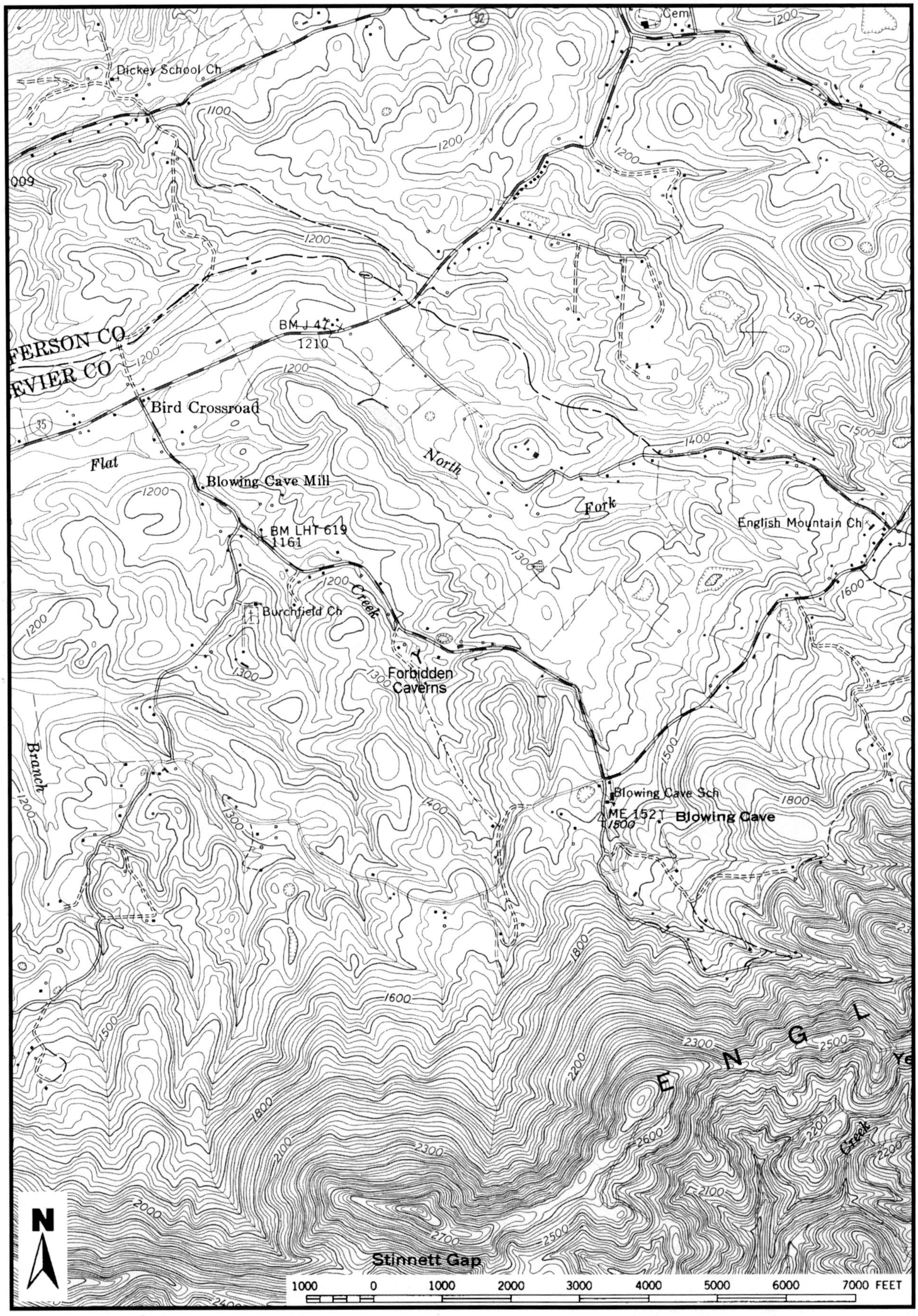

Figure 8.1. Topographic map showing the location of the entrance to Forbidden Caverns.

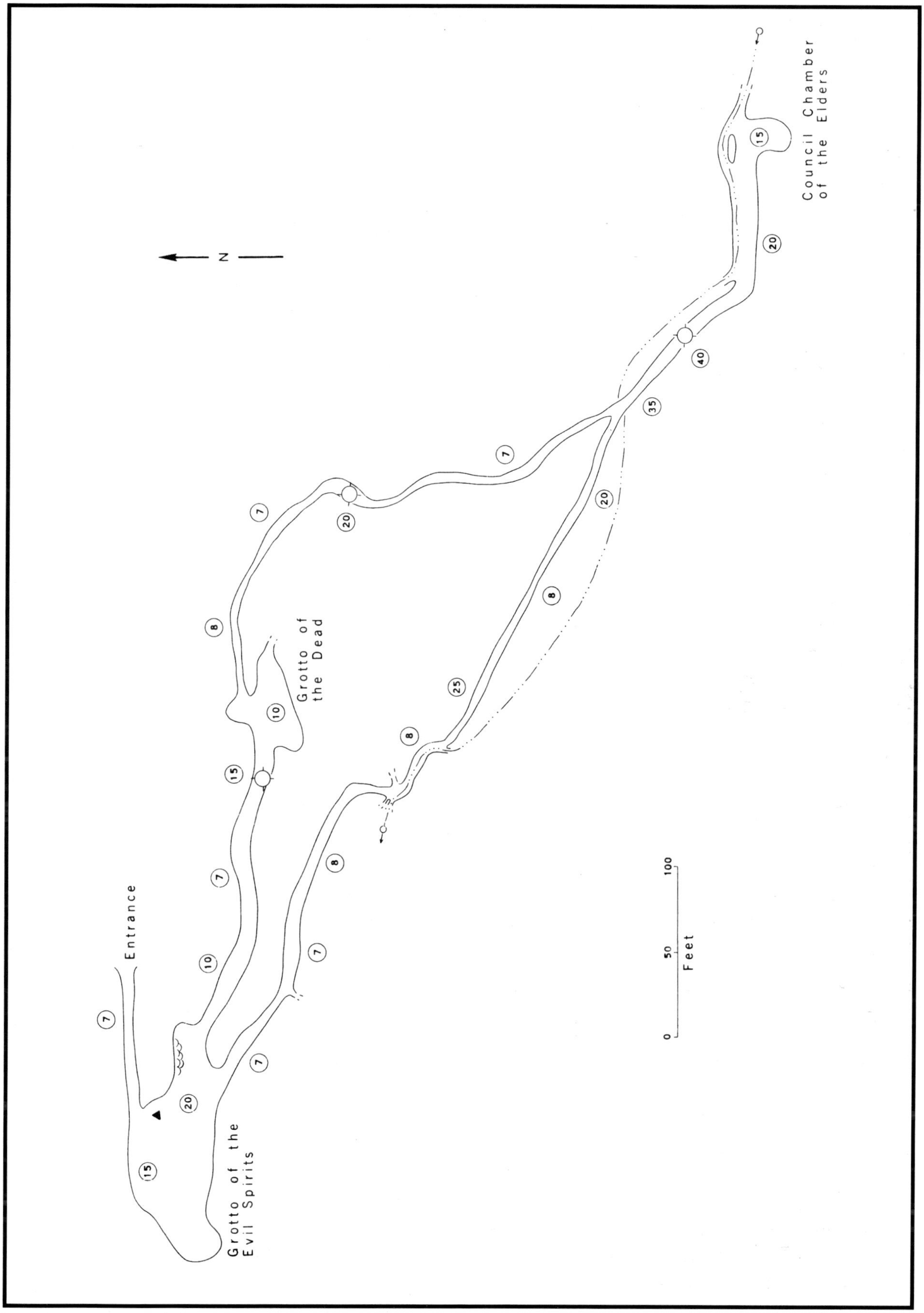

Figure 8.2. Map of Forbidden Caverns from Matthews' Descriptions of Tennessee Caves *(1971) page 84. Mapped by Larry E. Matthews, George L. Benedict III, and Richard M. Craig, 1968.*

Figure 8.3. Front and back of an early brochure for Forbidden Caverns. From the Speleo Archives of Gary K. Soule.

Forbidden Caverns has a surveyed length of 3,617 feet, which makes it the second longest known cave in Sevier County. Only nearby Cuyler Cave (described below) is longer.

Cuyler Cave

When you stand outside Forbidden Caverns, nearby English Mountain dominates the skyline. Underneath English Mountain lies Cuyler Cave, the longest known cave in east Tennessee. The original name given to this cave was Blowing Cave and Thomas C. Barr Jr, described it in his 1961 book, *Caves of Tennessee*. At that time, very little of the cave had been explored. Barr's description of the cave is as follows:

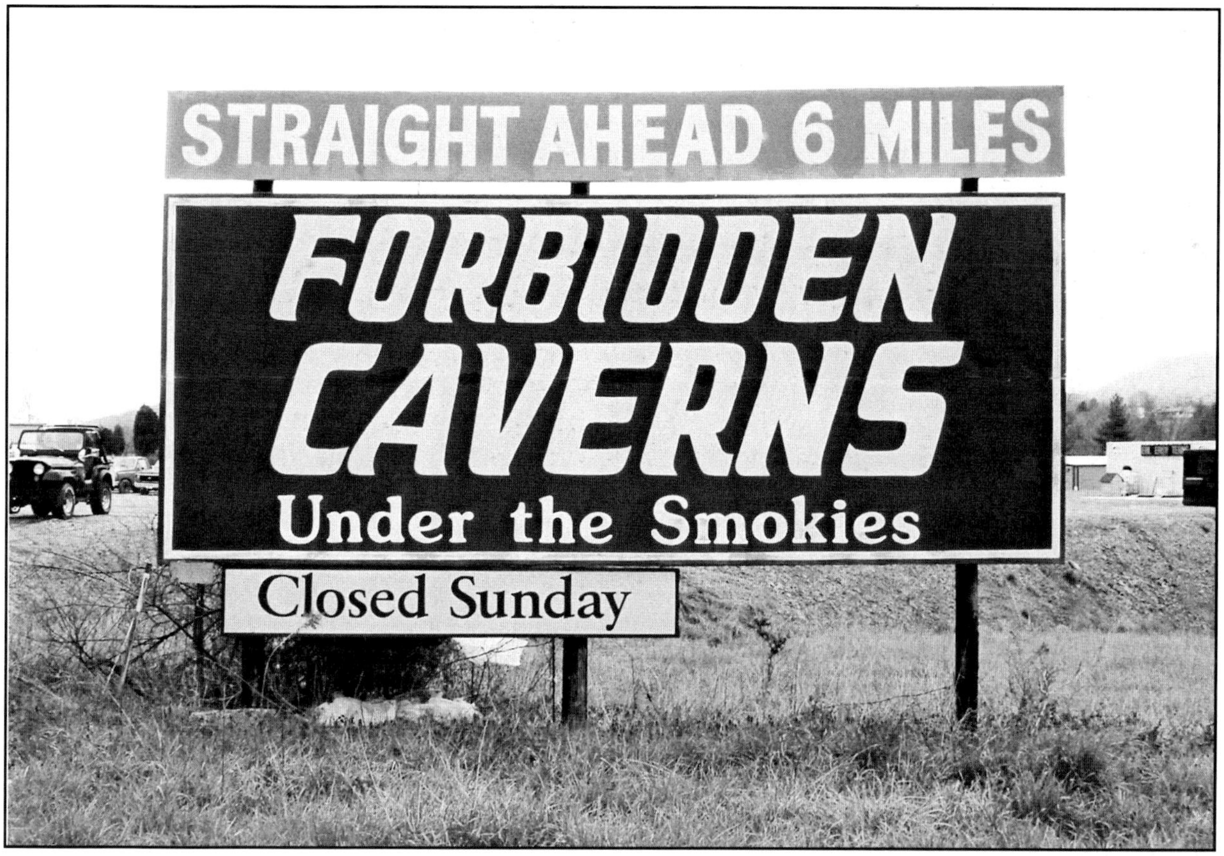

Figure 8.4. Billboard for Forbidden Caverns. Photo by Larry E. Matthews, April 2, 2008.

Figure 8.5. Larry Matthews next to the Forbidden Caverns welcome sign.
Photo by Jim Whidby, April 2, 2008.

Figure 8.6. Visitor center for Forbidden Caverns. Photo by Larry E. Matthews, April 2, 2008.

Figure 8.7. Jim Whidby in the entrance tunnel to Forbidden Caverns. Photo by Larry E. Matthews, April 2, 2008.

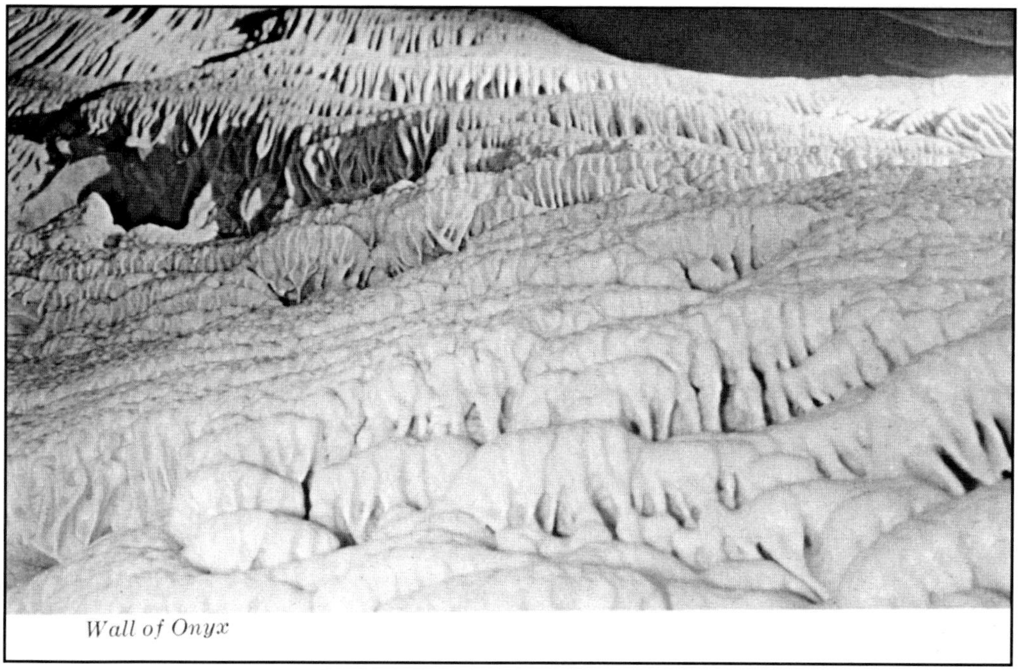

Figure 8.8. Postcard showing the "Wall of Onyx" in Forbidden Caverns. From the Speleo Archives of Gary K. Soule.

The mouth is a triangular hole 7 feet wide and 3½ feet high. A strong breeze issues from the cave in warm weather. The entrance room, the largest part of the cave, is about 60 feet in diameter and 8 feet high. A crawlway extends along the strike for 100 feet to another, much smaller room which contains a few dripstone formations. Some other crawlways extend downward around the edge of the entrance room but are penetrable only for short distances. The cave is rather dry.[2]

2 Thomas C. Barr Jr, *Caves of Tennessee*, Bulletin 64, Tennessee Division of Geology, 1961, p 414.

Figure 8.9. Postcard showing the "Council Chamber of the Elders" in Forbidden Caverns. From the Speleo Archives of Gary K. Soule.

Figure 8.10. Postcard showing the "Crystal Clear Stream" in Forbidden Caverns. From the Speleo Archives of Gary K. Soule.

Ten years later, Larry E. Matthews described Cuyler Cave in his 1971 book, *Descriptions of Tennessee Caves*. Matthews' description of the cave is as follows:

> The entrance is in the bottom of a small sink and opens into a large breakdown room. One of the crawlways extending from this room leads to an 80-foot pit which drops into the major portion of the cave. This lower level contains a number of very long stream passages. As of July 1970, the East Tennessee and Smoky Mountain Grottos of the National Speleological Society had mapped 6.7 miles of passages, making this undoubtedly the largest known cave in east Tennessee and one of the very largest caves in the entire state.

Figure 8.11. Postcard showing the "Big Rock Candy Mountain" in Forbidden Caverns. From the Speleo Archives of Gary K. Soule.

Figure 8.12. Postcard showing the "Frozen Waterfall" in Forbidden Caverns. From the Speleo Archives of Gary K. Soule.

Figure 8.13 Postcard showing the "Valley of the Moon" in Forbidden Caverns. From the Speleo Archives of Gary K. Soule.

Figure 8.14. Jim Whidby next to a large stalactite in Forbidden Caverns. Photo by Larry E. Matthews, April 2, 2008.

> Forbidden Caverns is believed to be the downstream end of the Cuyler Cave stream system, although it may not be possible to make a connection between the two caves.[3]

Since Matthews (1971) describes this as a separate cave, it seems likely that the connection to Blowing Cave had not been made at that time. However, the two caves have now been connected and the Blowing Cave Entrance and the Cuyler Cave Entrance both enter into the same cave system, Cuyler Cave. The surveyed length of Cuyler Cave is 8.16 miles (13.1 kilometers), making this the longest known cave in east Tennessee and the 10th longest known cave in the state of Tennessee.

Studies have shown that the stream in Cuyler Cave flows northwest into Forbidden Caverns. After flowing through Forbidden Caverns, the water emerges as a spring in an abandoned quarry just a few hundred yards north of Forbidden Caverns. From a hydrological standpoint, Cuyler Cave and Forbidden Caverns are both part of a vast underground drainage system that drains English Mountain. But to cavers, they remain separate caves until someone discovers a passage that connects the two caves.

Cuyler Cave is not open to the general public and both entrances are securely gated.

Forbidden Caverns Today

Forbidden Caverns is open to the public from April 1 through November 30. It is closed on Sundays. The cave property is located at 455 Blowing Cave Road, near Sevierville, Tennessee.

The average guided tour last 55 minutes. Cave temperature is 58 degrees year-round, so a jacket or light sweater is recommended. As in

3 Matthews, op. cit., pp 82–83.

Figure 8.15. Jim Whidby by the remains of a moonshine still on display inside Forbidden Caverns. Photo by Larry E. Matthews, April 2, 2008.

Figure 8.16. Tour guide next to old barrel hoops left from a moonshine still. Photo by Jim Whidby, April 2, 2008.

Figure 8.17. Jim Whidby next to the Gargoyle in Forbidden Caverns.
Photo by Larry E. Matthews, April 2, 2008.

all caves, sturdy footwear is recommended.

For more information on Forbidden Caverns, telephone:

865-453-5972

You can check out their Web site at:

http://www.forbiddencavern.com

Their mailing address is:

Forbidden Caverns
455 Blowing Cave Rd
Sevierville TN 37876-1270

Directions

From Knoxville, go east on I-40 for 20 miles to Exit 407. Turn right (south) on Highway 66. Go 8 miles to Sevierville, then turn left (east) on U.S. 411. Go 10 miles and turn right on Blowing Cave Road. The entrance to Forbidden Caverns is 2 miles ahead and on your right. There are numerous signs all the way from the Interstate, so Forbidden Caverns is very easy to find.

Figure 8.18. Close-up of an eastern pipistrelle bat in Forbidden Caverns. Photo by Jim Whidby, April 2, 2008.

Figure 8.19. Old grist mill near Forbidden Caverns. Photo by Jim Whidby, April 2, 2008.

Figure 8.20. Front and back of a recent brochure for Forbidden Caverns.

Figure 8.21. Forbidden Caverns sticker. From the Speleo Archives of Gary K. Soule.

Figure 8.22. Larry Matthews next to flowstone formations. Photo by Jim Whidby, April 2, 2008.

Figure 8.23. Forbidden Caverns matchbook. From the Speleo Archives of Gary K. Soule.

Side Trips

After driving down Blowing Cave Road for 0.4 mile, you will come to an old mill on your left. Currently, this mill is open to tourists and is well worth a visit.

Just before you reach Forbidden Caverns, you will pass a trout farm and restaurant on your right. Here, you can have the exciting experience of catching your own trout and having them cooked and served to you just minutes later.

If you are staying in Gatlinburg or Pigeon Forge, when you return to U.S. 411, turn right and go 7 miles to Exit 423 of I-40. Turn right onto I-40 and go east for 20 miles to Exit 443. This is the exit for the Foothills Parkway. The Foothills Parkway extends southwest for 5 miles and provides breathtaking views of the Great

Figure 8.24. Sign for the English Mountain Trout Farm. Photo by Larry E. Matthews, April 2, 2008.

Figure 8.25. Forbidden Caverns patch. From the Speleo Archives of Gary K. Soule.

Smoky Mountains to the southeast. Turn left on U.S. 321 and follow this highway west for 20 miles to Gatlinburg. For the entire drive, the Great Smoky Mountains National Park will be to your left (south) and the scenery is magnificent.

To return to Knoxville from Gatlinburg, drive north on U.S. 441 to Sevierville. In Sevierville, continue straight on Highway 66 and return to Exit 407 of I-40. Turn left (west) onto I-40 and return to Knoxville.

Chapter 9

Gregorys Cave

The Discovery of Gregorys Cave

Gregorys Cave is located on the north side of Cades Cove at the base of Cave Ridge. The first white settler in Cades Cove was John Oliver (1793–1863), who arrived in 1818. Gregorys Cave is located 3,200 feet west of the original John Oliver Cabin. Since the early settlers relied heavily on hunting and gathering fruits and nuts for food, Gregorys Cave would have been discovered by John Oliver and his family, or other settlers, very soon after they arrived. Clearly, the cave would have had a different name during this time period, but that name is not recorded.

Saltpeter Mining

During a research trip into Gregorys Cave on May 17, 2008, the author observed tally marks on the south wall of the cave just past the side passage that leads to the south. The dirt deposits on this side of the main cave passage in this area show evidence of mining and pick marks are visible in the clay dirt. Since these excavated dirt faces are under ledges or in areas of low ceiling height, they are relatively undisturbed and the evidence is well preserved. The side passage itself may also have been mined, but so many people have entered this passage, that any evidence has been erased. Tally marks and the removal of dirt would indicate that Gregorys Cave was mined for saltpeter sometime between 1818, when Cades Cove was first settled, and 1865, the close of the Civil War. It is likely that the saltpeter leaching vat(s) would have been located outside of the cave and that the dirt removed from the cave would have been processed there. There is no evidence inside the cave of any leaching vat(s) at the present time. Due to the relatively small amount of dirt that was available in this cave, this would have been a small mining operation.

The map by Jonathan Mays (Figure 9.3) shows a crawlway, also on the south side of the main cave passage and two hundred feet past the entrance, that is labeled "Clay Mining Crawlway" (See Figure 9.3.). This may have also been a part of the saltpeter dirt mining operation. However, Mays states that, "Gregory also profited from the cave by selling clay for the manufacture of tobacco pipes and giving tours for a small fee." It seems likely that the majority of the dirt removed from the cave was the result of saltpeter mining and that perhaps a small amount of clay was removed later to manufacture pipes.

The History of Gregorys Cave

Russell Gregory (1795–1864) was living in Cades Cove by 1838. However, there is no record of the year when the portion of land on which the cave is located passed into the Gregory family. However, we do know that the property eventually ended up in the hands of Josiah Jonathan ("JJ") Gregory (1870–1933). "JJ" was the great-grandson of Russell Gregory.[1] The cave was used by the Gregory family to store food, in the days before refrigeration.[2] Eventually, the Gregory family developed the cave commercially and gave guided tours. A property assessment card on file at the Great

1 A. Randolph Shields, *The Families of Cades Cove, 1821–1936.*

2 Anonymous, "Philadelphia Grotto Holds Well Attended Summer Meeting," *NSS News*, vol 5 (1947), no. 9, p 2.

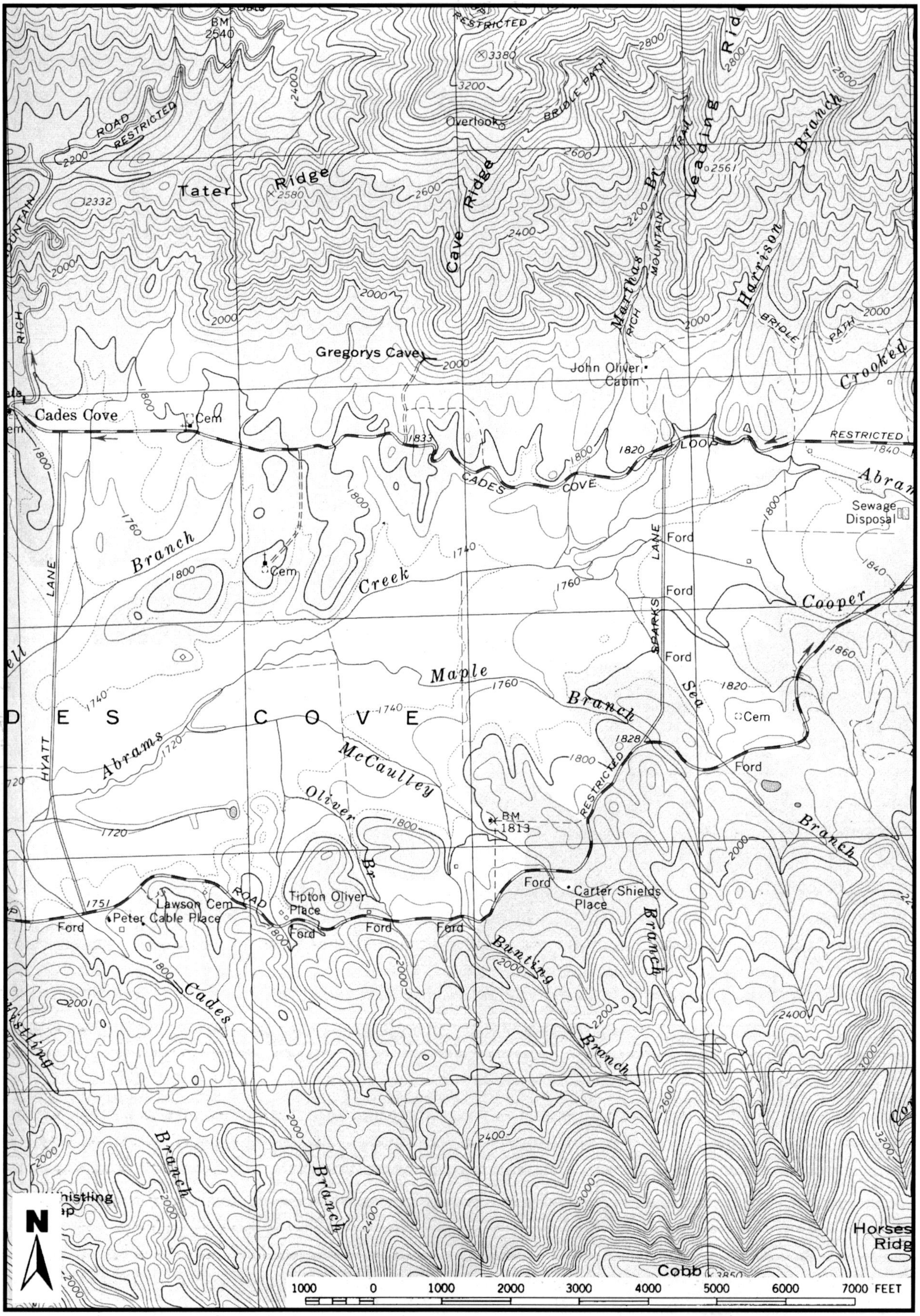

Figure 9.1. Topographic map showing the location of the entrance to Gregorys Cave.

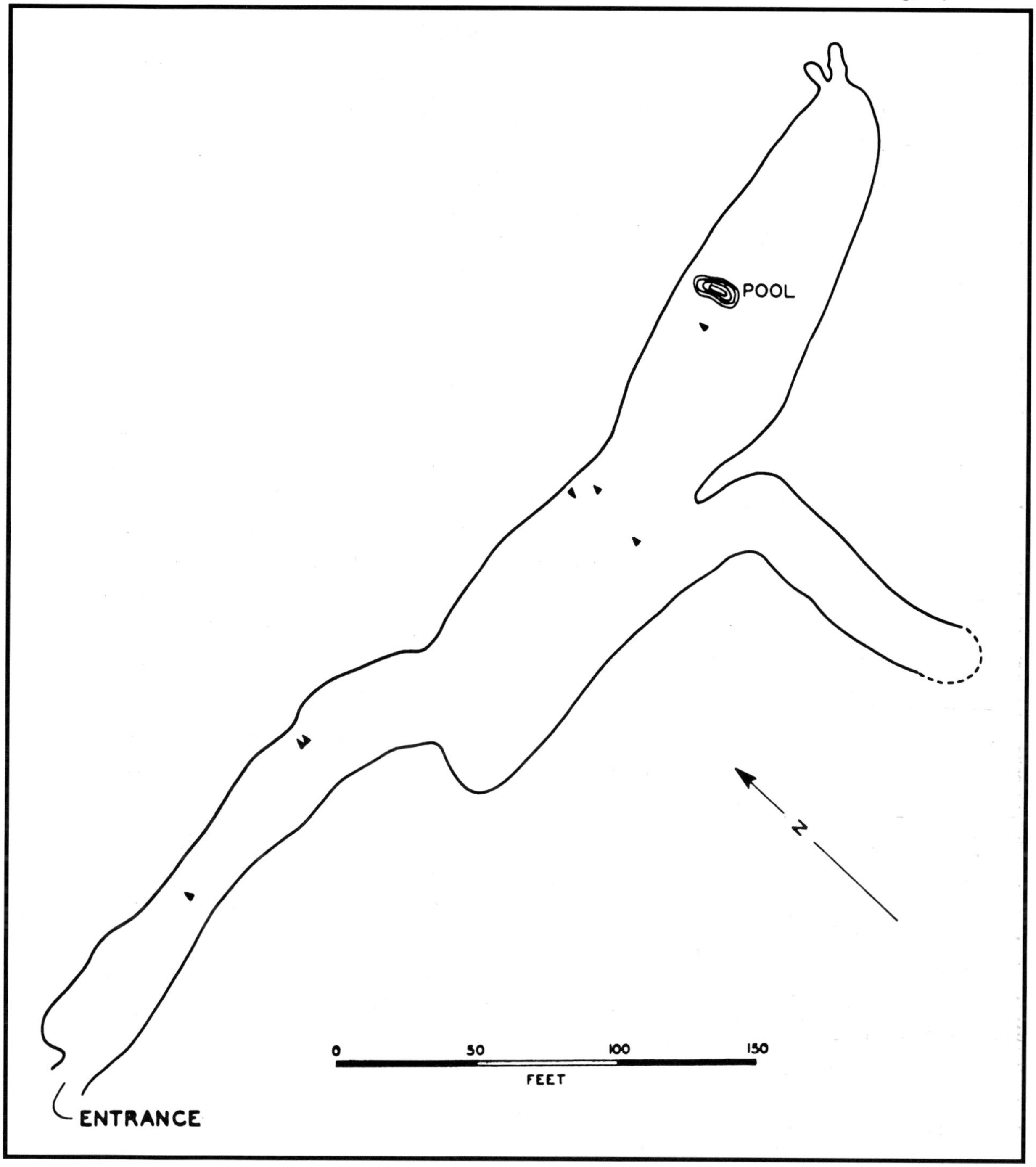

Figure 9.2. 1929 map of Gregorys Cave from Barr's Caves of Tennessee *(1961), page 78. Survey by Berlen C. Moneymaker.*

Smoky Mountains National Park Archives in the Sugarlands Visitor Center gives the following information:

> The owner of the property is J.J. Gregory. The address is Cades Cove in Blount County and the size of the property is 114.4 acres. The description is dated 4/7/31 and is as follows: "40 acres fair and poor cultivated land and 74.4 acres in good oak-pine timber. Dirt road and wire-rail fence. 5-room frame house in good condition. 6 other buildings: barn, 2 residences, smokehouse, crib. Improvements $1,800 to $2,000." Under remarks it states: "Improvements about $1,800 to $2,000. This farm has large cave electrically lighted and produces income of $400 to $600 annually."

The assessor valued the property

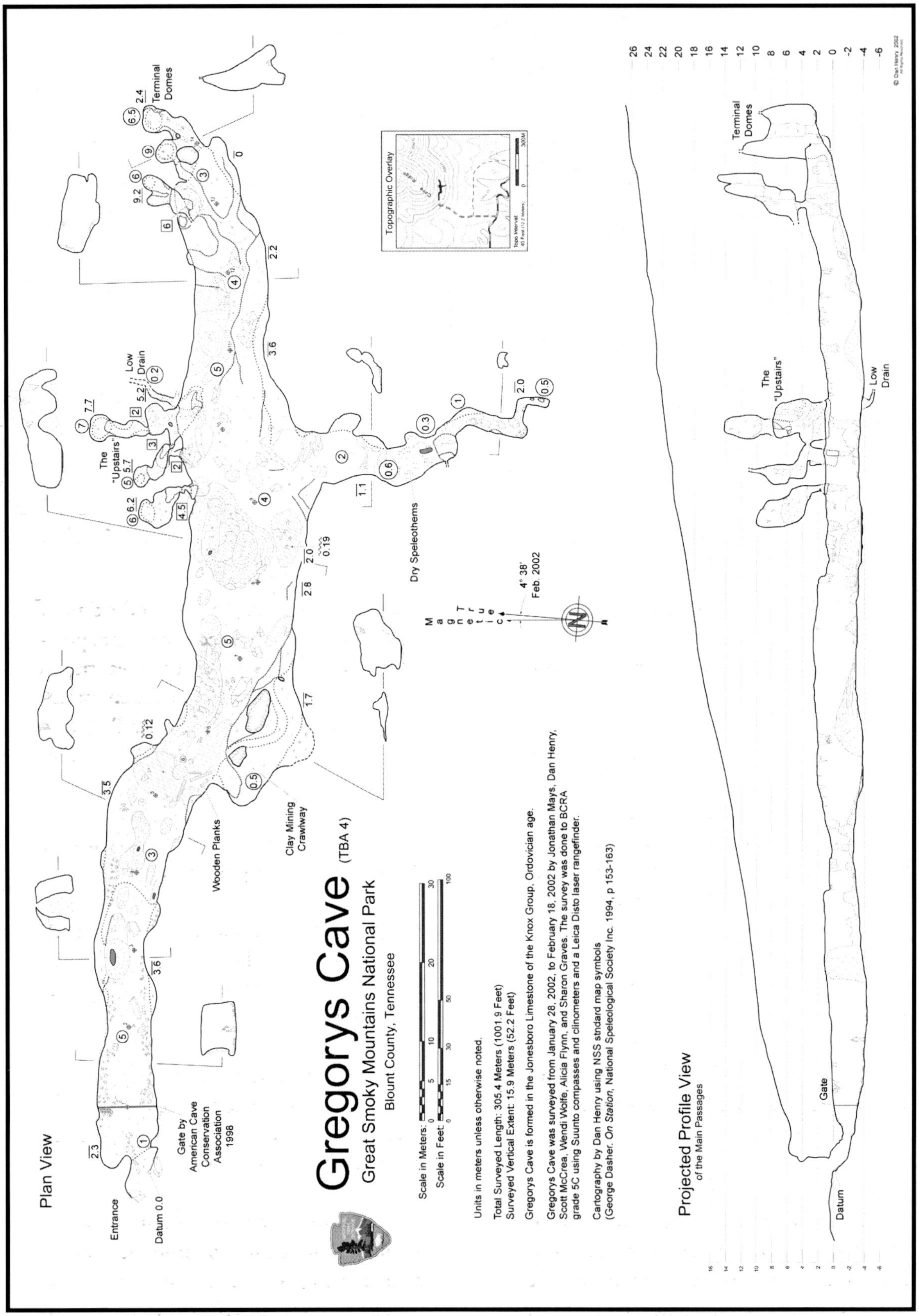

Figure 9.3. 2002 map of Gregorys Cave by Dan Henry and others. © 2002 Dan Henry, used by permission.

Figure 9.4. Photograph of the Gregory House, 1954. Courtesy of Great Smoky Mountains National Park.

> at $10,000, but the owner was asking $25,000. Since they could not agree on a price, the condemnation sale was contested in court and the court agreed with the $10,000 valuation. According to the property assessment card, the Final Decree was received on 11/13/33, the decree was recorded and returned on 11/24/33 and the deed was delivered to the state on 5/7/35. The state then transferred to deed to the federal government on 5/13/35.

Another report on file at the park library is titled *Geology of The Great Smoky Mountains National Park* by Donald K. MacKay, Geologist, National Park Service. This report is dated September 14, 1935. Mr MacKay makes the following comments:

> Gregorys Cave has been open to the public since July 1925. Adults are charged 50 cents and children are admitted free. The cavern was formerly used by the Gregory families as a storehouse for fruits, vegetables, and other perishable foods, and the temperature is said to be between 40° and 46° throughout the year. At the time of this visit, the rather agreeable odor of potatoes pervaded the cave, and this was traced to a small pile of tubers near the entrance.
>
> The writer believes that the Park Service should take over the operation of Gregorys Cave, in order to:
>
> (a) Insure the preservation of the cave deposits in their original state,
>
> (b) Provide standard conveniences, such as proper lighting and drainage, which at the time of this visit were not altogether satisfactory.

When MacKay visited the cave, it was still being shown commercially. Here is his account of that tour and his observations of the cave:

> The cave is the result of solution along a joint in the Knox Dolomite. It is an irregular elongated cavern having an east to west trend. It is 590 feet long (the total cave length is approximately 700 feet), and is from 30 to 60 feet wide, the average

Figure 9.5. Photograph of J.J. Gregory playing the fiddle for his grandson. Courtesy of Great Smoky Mountains National Park.

width being 43 feet. The height of the roof above the floor varies from 8 to 30 feet, the average being about 19 feet. The floor is approximately 6 feet below the level of the surface at the entrance, and is covered with moist clay. During wet weather, pools of water occur in the floor at various places but there is only one permanent pool and this is located near the far end of the cavern.

Stalactites and stalagmites of various sizes occur thickly and some of the stalagmites are unusually large, having diameters of from 10 to 12 feet. In the floor around some of these there have developed natural rock-rimmed basins which contain water in rainy weather. The stalactites and wall deposits are numerous and in the dim electric light available, appear beautiful. When stalactites and stalagmites have grown together they form large pillars. Some of the wall deposits are covered with fine wart-like deposits. Many peculiar shapes are seen in the cave deposits and these are perhaps more strikingly shown because of the dim light. The guide pointed out shapes bearing resemblances to: a bear, snake, turtle, and baseball glove, and enlivened the trip with imaginative stories concerning these and other shapes whose resemblances were not so apparent.

National Speleological Society members Charles Mohr and Ernest Ackerly, along with Assistant Park Naturalist Dick Burns, visited Gregorys Cave on Monday, December 4, 1950, where they reported finding a small colony of long-eared bats and "a variety of other forms of cave life." It is interesting to note that they refer

Figure 9.6. Erica Sughrue examines tally marks left by saltpeter miners in Gregorys Cave. Photo by Bob Biddix, May 17, 2008.

Figure 9.7. Pick marks left in the clay dirt by saltpeter miners. Photo by Bob Biddix, May 17, 2008.

Figure 9.8. Pine torches left by early explorers in Gregorys Cave.
Photo by Bob Biddix, May 17, 2008.

to Cades Cove as "seldom-visited."[3] Today, Cades Cove is one of the most frequently visited portions of the Park.

Caver Graham Ward provides some additional information on the commercial history of Gregorys Cave:

> Gregorys Cave is located 200 yards to the right of the Loop Road about one mile from the Cades Cove section of the Great Smoky Mountains National Park. The cave gets its name from Jo Gregory who was the land owner prior to park acquisition. It is found in Knox Dolomite of Ordovician age which interestingly is capped by interbedded Cambrian sandstone and shale due to an overthrust fault during the Appalachian Revolution.
>
> The cave is roughly 500 to 600 feet long and consists of one main, nearly-straight gallery about 30 feet wide and high. A smaller passage at the end of the main gallery continues for about 20 feet and ends at the bottom of a domepit. The entire length is well decorated. There are a number of very large stalagmites and many rimstone terraces plus a large assortment of stalactites, wall draperies, and other flowstone deposits. Much of the ceiling and higher formations have been blackened by smoke. Mr Shields, the ranger naturalist who accompanied us and [is] a naturalist of the area, said that as a boy he and other local youths often visited the cave using pine knot torches. When the cave was first taken over by the Park Service, it was wired for lights and a few rough walkways were

3 Anonymous, "NSS Makes Cave Survey In Great Smoky Mt. National Park," *NSS News*, vol 9 (1951), no. 1, pp 1, 4.

> built over the more muddy sections, but the lights have since been removed and the walks are in a bad state of repair. Only one bat was seen, but since it was flying the identity is uncertain. It was probably a little brown bat.[4]

In the 1951 *NSS News* article cited earlier, another nearby cave that had been shown commercially is mentioned:

> Exploration of Great Smoky Mountain Cavern on Sunday and its mapping later, discounted rumors of a connection between the two caves. The lower, quite spectacular 2,200 feet of this cave was mapped on December 5 (1950) by Ackerly, Nelson, and Jerry Hadley.
>
> Part of this cave had been open to tourists in the 1930s. A 30-foot stairway still stands, in highly dangerous condition. Efforts to remove this hazard were unsuccessful. A steel ladder was used to make the descent.[5]

The "two caves" that they are referring to are Whiteoak Sink Cave inside Great Smoky Mountains National Park and Great Smoky Mountain Cavern, located outside the Park. The name currently in use for what they referred to as Whiteoak Sink Cave is Whiteoak Blowhole and it is described later in this chapter.

Great Smoky Mountain Cavern was shown commercially during the early 1930s. This cave is now known as Tuckaleechee Caverns and it is described in Chapter 11 of this book. Tuckaleechee Caverns is approximately 2 miles north-northwest of Whiteoak Sink. Barr states that:

> A sizeable stream flows through the cave [Tuckaleechee Caverns], emerging from a siphon upstream and sinking beneath the floor at the downstream end. The stream is thought to originate in White Oak Sink, 2 miles southeast of the caverns' entrance, and probably emerges at Dunn Spring, ¾ mile northwest of the entrance.[6]

A Voice From the Past

In an effort to record the cultural history of the Great Smoky Mountains, persons who had lived inside the Park were interviewed and the conversations recorded on cassette tapes. The following is a transcript of a conversation between Wesley Herman Gregory (born 1906), the son of J.J. Gregory, and Lloyd Foster of the Tremont Environmental Education Center. This interview took place on April 22, 1971. Only the portions of the recording pertaining to Gregorys Cave are reproduced below.

> Gregory: Then we's first talkin' 'bout the cave that there's right on the old homeplace. Way back fur as I can remember, just Cades Cove folks'd come in there, young people an' go in there Sunday afternoons you know. See it's always cool in there. It's 45 and 46 degrees the year round. And. But if you go in there and then rush out I've knowed of people a fainting. Rush out into the heat say in July or August. You should take your time a coming out of there when it's real hot.
>
> Foster: When did your family stop living up there?
>
> Gregory: In 1943. The fall of 1943. My mother, her and Uncle Noey Burchfield was the only two people that I know of in the Cove got a whatcha call a lifetime dowry take it wouldn't be put out til they passed away. And she passed away March 26, 1943, and my brother lived with her, course he finished his crop that year. Then

4 Graham Ward, "Gregory's Cave (Great Smoky Mt. Natl. Park, Tennessee)," *1958 Speleo Digest*, p 1-242.

5 Anonymous, "NSS Makes Cave Survey In Great Smoky Mt. National Park," *NSS News*, vol 9 (1951), no. 1, pp 1, 4.

6 Thomas C. Barr Jr, *Caves of Tennessee*, Bulletin 64, Tennessee Division of Geology, 1961, p 82.

he moved out.

Foster: I guess that land was cultivated after that for a spell wasn't it?

Gregory: No I don't think that was. It was put back in pasture the most of that. You see that there's. My old homeplace was, well I don't guess you can even find it now. The best land was on up next to the mountain 'round to the left. And but that right around the old homeplace there was hilly. Course they was good garden out through there. Always had a garden, 'bout a acre of it.

Foster: Did–when was it they had electric lights strung in there? In the cave.

Gregory: Well that there was after, that's after I moved to Knoxville but I was back up there a lot. It's after 1925 sometime. And. Governor of Tennessee went through there, and people from London, England. I've got a book of that people that 'd sign an' go through there. It was a Delco system. It. And course they they're damp places in there that you always had to put some walks you know had to put walks and I forget what else.

Unidentified Teacher: When the lights were in the cave was there also a wooden like sidewalk that was built from the entrance on back?

Gregory: Not from the entrance, on back a little where it stayed muddy all the time where it was so damp. Right there where you go in the first was about the biggest widest place they are you know of the cave.

Mrs. W.H. Gregory: They did have, you know, food stored in there. Is there anything in there?

Foster: Yeah there still is.

Unidentified Teacher: A fallout shelter.

Gregory: Have they got it?

Unidentified Student: Was there an entrance running back of the cave? I mean leading out, an exit?

Gregory: What did he say?

Unidentified Teacher: Is there another entrance or exit?

Gregory: No there ain't none not as I ever knowed of. 'N' I've been in there hundreds of times and looked. 'N' Right there when you first go in, now I never was too much to crawl around in holes as wasn't any size. You could go what you called upstairs. Some go up there. I don't know where it went but it didn't go out. Anywhere.

Foster: Mr Gregory what was it you were going to ask there about that stockpile? You were going to ask something about the stockpile up there, the fallout shelter.

Gregory: Oh! Is it stocked now?

Foster: Yeah it still is. As a matter of fact they were up there about two years ago and replenished that stock.

Gregory: Yes. Well we used to keep potatoes in there and keep 'em from fall til time to whatcha call sweet potatoes to beddin' you know and, git you sprouts to set out.

Unidentified Teacher: Did you ever play any games in the cave as a boy you know?

Gregory: Oh yeah. They used to pick the banjo and play the violins and all such as that but a banjo, it would, wouldn't sound right for a little while that dampness would make that head you know a banjo head is made out of leather. And it would flop down til it wouldn't sound too good but you take music sounds wonderful in there.

Unidentified Student: Did you ever try to catch you a bat in that cave?

Gregory: I've caught 'em. Lots of 'em. And. I don't know whether I should tell that or not. Now I haven't done it personally but I've seen it done. Cut a little hole in that there bat a take a quill and blow it up and make it a great big bat.

The complete transcript of the above interview is on file at the Great Smoky Mountains National Park Archives, located in the Sugarlands Visitor Center. This conversation shows that the Gregorys continued to live on their farm even after the deed passed to the Park Service in 1935. W.H. Gregory's mother died on March 26, 1943, and then his brother moved out later in the year. This transcript also proves that the cave was stocked and still being used as a fallout shelter in 1971. It also documents the existence of wooden walkways and electric lights in the cave. The lighting system is referred to as a "Delco" system. This was a Delco generating system that has been used in several caves, including Diamond Caverns in Kentucky.

A Description of the Cave

Gregorys Cave is not open to the public at the present time. The book, *Caves of Tennessee*, which was written by Thomas C. Barr Jr, in 1961, gives the following written description of the cave:

> Gregorys Cave extends eastward 435 feet as a gallery 20 to 55 feet wide. The average height is 15 feet. A low, wide side passage extends southward for 120 feet. Stalactites, stalagmites, and rimstone are well developed, and several pools are present. The cave was formerly commercialized and contains the rotted remains of walkways and bridges. The floor is remarkably level throughout.[7]

7 Barr, op. cit., pp 77–78.

A 1929 map of Gregorys Cave, by Berlen C. Moneymaker, is reproduced in Figure 9.2. Jonathan Mays and others prepared a modern map, with significantly more detail, in 2002. This map was drawn by Dan Henry and is reproduced in Figure 9.3.

The Life in Gregorys Cave

Gregorys Cave contains two salamanders: *Eurycea longicauda* and *Plethodon glutinosus. Eurycea longicauda* is a beautiful orange salamander with black spots. It is frequently found at cave entrances and inside caves. It is commonly known as the "Long-tailed Salamander." *Plethodon glutinosus* is known as the "Slimy Salamander" and is not a true cave dweller, or troglobite. It is occasionally found at cave entrances, however. Nathan Taylor and Jonathan Mays studied the salamanders of Gregorys

Figure 9.9. The northern slimy salamander, Plethodon glutinosus, *in Gregorys Cave. Photo by Jonathan Mays.*

Figure 9.10. The long-tailed salamander, Eurycea longicauda, *in Gregorys Cave. Photo by Jonathan Mays.*

Figure 9.11. The cave cricket, Euhadenoecus *sp., in Gregorys Cave. Photo by Jonathan Mays.*

Cave, making monthly visits for a year. During this time period they observed both adults and juvenile salamanders in the cave. The salamanders were observed from near the twilight zone to near the end of the main passage. Other amphibians were occasionally observed in the cave, including American toads.[8]

Gregorys Cave contains 46 arthropod species representing 5 classes, 16 orders, and 35 families. These arthropods were studied over a 12-month time period by Jonathan D. Mays, from May 2000 to April 2001. Among the various arthropods found in Gregorys Cave were amphipods and isopods, which were observed living in rimstone pools. Arthropods found living outside the pools included crickets, flies, millipedes, and spiders. Two exciting discoveries during this study were a new species of troglobitic millipede, *Scoterpes* sp. n., and a new species of springtail, *Pseudosinella* sp. n.

Troglobites are animals that are so specialized that they can only live in caves. All three crustaceans that live in Gregorys Cave are classified as troglobites. They are:

Stygobromus fecundus - An amphipod.

Stygobromus sparsus - An amphipod.

Caecidotea incurva - An isopod.

Of the eight arachnids that live in Gregorys Cave, only one is classified as a troglobite:

Nesticus barrowsi - A spider.

Out of six diplopods, one is classified as a troglobite:

Scoterpes sp. n. - A millipede.

The remainder of the animals found in Gregorys Cave were classified as trogloxenes (11), troglophiles (16), accidentals (5), and unknowns (9). **Trogloxenes** are animals that spend part of their life underground and part on the surface. **Troglophiles** are animals that can live in caves, but also live in cool, moist habitats on the Earth's surface.[9]

Figure 9.12. The cave cobweb spider, Nesticus barrowsi, *in Gregorys Cave. Photo by Jonathan Mays.*

The Cold War Comes to Cades Cove

One of the items in the Park Library is an undated newspaper clipping, titled the "Times Action Line." It reads as follows:

Q. Someone told me that there are two

8 Nathan Taylor and Jonathan Mays, "The Salamanders *Eurycea longicauda* and *Plethodon glutinosus* in Gregorys Cave, Tenn: Monitoring and Observations on Ecology and Natural History," *Southeastern Naturalist*, vol 5 (2006), no. 3, pp 435–442.

9 Jonathan D. Mays, *A Systematic Approach to Sampling the Arthropod Assemblage of Gregorys Cave, Great Smoky Mountains National Park*, Masters Thesis, Western Carolina University, April 2002, 54 pages.

Figure 9.13. Erica Sughrue at the entrance to Gregorys Cave. Photo by Bob Biddix, May 17, 2008.

Figure 9.14. Erica Sughrue looking at the inside of the Gregorys Cave gate. Photo by Bob Biddix, May 17, 2008.

Figure 9.15. Erica Sughrue by flowstone with rimstone dams in Gregorys Cave. Photo by Bob Biddix, May 17, 2008.

caves in the county that are fallout shelters. I am guessing that Tuckaleechee Caverns is one of them but I can't think of another large cave in the area that might be suitable for this purpose.

A. You are right about Tuckaleechee Caverns. Civil Defense officials list it as a shelter with a capacity of 3,500. The other cave listed among the 29 facilities is Gregorys Cave in Cades Cove. It has a capacity of 1,000.

During the height of the Cold War, the federal government not only considered caves for use as fallout shelters, they actually went so far as to designate a number of caves as official public fallout shelters. A few of these caves were actually stocked with supplies, and Gregorys Cave was one of those that were stocked. The theory was that if people had enough warning, they could go to one of these designated caves and be protected, first from the blast of the atomic bomb when it exploded, and then spend several more weeks protected from the radioactive fallout. In retrospect, it is doubtful that people would have had adequate warning before a nuclear attack to seek shelter in these caves. Further studies of caves showed that the considerable flow of air and water through their passages would have resulted in their being of little, if any, value for protection from fallout.

To further complicate matters, caves are relatively cold and damp. Fifty-six degrees seems delightful when you are caving, but if you sit down for more than ten minutes, you definitely start to get cold. For people to spend days or weeks in a cave, they would need to be dressed much more warmly than people in the South are accustomed to dressing. And, like most caves, Gregorys Cave is muddy. Where would people sit and sleep? What provisions would be made for restroom facilities? What would be the source of light? All of these problems appear to have been ignored in the effort to make the populace believe that they actually had a safe place to go in case of nuclear war.

The assigned capacity for Gregorys Cave of 1,000 people is particularly amazing. The cave contains approximately 20,000 square feet of passage, so each person would have only 20 square feet of floor space, or an area less than 5 feet square. Not a pleasant prospect for several weeks in the dark and cold.

Gregorys Cave Today

Gregorys Cave is securely gated and it is closed to the public. To the best of the writer's knowledge, it is impossible to obtain a permit to enter this cave unless you are conducting some sort of scientific research.

It is worth your time, however, to park along the Cades Cove Loop Road and walk the 1,200-foot-long trail from the road to the cave entrance. Along the way, you will pass the foundation of the Gregory's home. At the time this land was purchased by the Park Commission, 40 acres were cleared farmland that is now slowly reverting to forest. The cave entrance is a scenic spot to sit, relax, and enjoy the beauty of the Great Smoky Mountains.

A Description of Bull Cave

Located approximately 2.4 miles due north of Gregorys Cave is Bull Cave. Bull Cave is located 750 feet southwest of Rich Mountain Gap, inside the boundaries of the Great Smoky Mountains National Park. Although Bull Cave has never been a commercial cave, it is mentioned here since it is the deepest known cave in Tennessee, with a total depth of 924 feet. It is also a significantly long cave, with 2.3 miles of surveyed passages. Caver Charles Clark reports that local residents named the cave when a bull fell down the steep entrance sink and died in the cave entrance.[10] The following description of Bull Cave is given in Barr's *Caves of Tennessee*:

> Bull Cave is one of the deepest vertical caves in the eastern United States. The mouth is a high, shallow opening at the base of a limestone bluff at the southeast side of a uvala 1,500 feet long, 500 feet wide, and 150 feet deep. The entrance room is 30 feet high, 85 feet long, and 25 to 50 feet wide. A small hole in the floor drops 10 feet to the top of a narrow fissure, the floor of which slants downward over breakdown at a steep angle. At 270 feet from the entrance is a 45-foot pitch, followed by a 50-foot climb down over rubble, to the top of a 150-foot pitch. The stream that sinks in the floor of the entrance room reappears and cascades down into a chasm nearby.
>
> The 150-foot drop opens into a vast breakdown chamber 100 feet high, 30 to 85 feet wide, and 260 feet long. A hole in the floor drops 30 feet into a winding gallery 20 feet high and 10 to 25 feet wide, the floor of which slopes steeply downward over breakdown; the stream flows from ledge to ledge, forming pools. At a distance of 400 feet from the big room, and about 135 feet below it, is a 105-foot drop down through a waterfall to the lowest level of the cave. At the bottom the stream strikes northeast through a high, narrow gallery and dips under a ledge 350 feet from the base of the falls. The end of the cave is 680 feet below the mouth, at an approximate elevation of 1,160 feet.
>
> The description of Bull Cave was provided by William F. Cuddington and John H. Gibbons, who, with members of the East Tennessee chapter of the National Speleological Society, explored the cave in July 1955. The accompanying map was drawn by Dr Gibbons.[11]

The 1955 map of Bull Cave, by J.H. Gibbons and T.C. McDaniels, is reproduced in Figure 9.16. A 1955 profile of the cave, comparing its depth to the Washington Monument, is reproduced in Figure 9.17. Like all caves in the Great Smoky Moun-

10 Charles Clark, "Bull Cave," *Journal of Spelean History*, vol 18 (1984), no. 3 & 4, pp 75–80.

11 Barr, op. cit., pp 76–77.

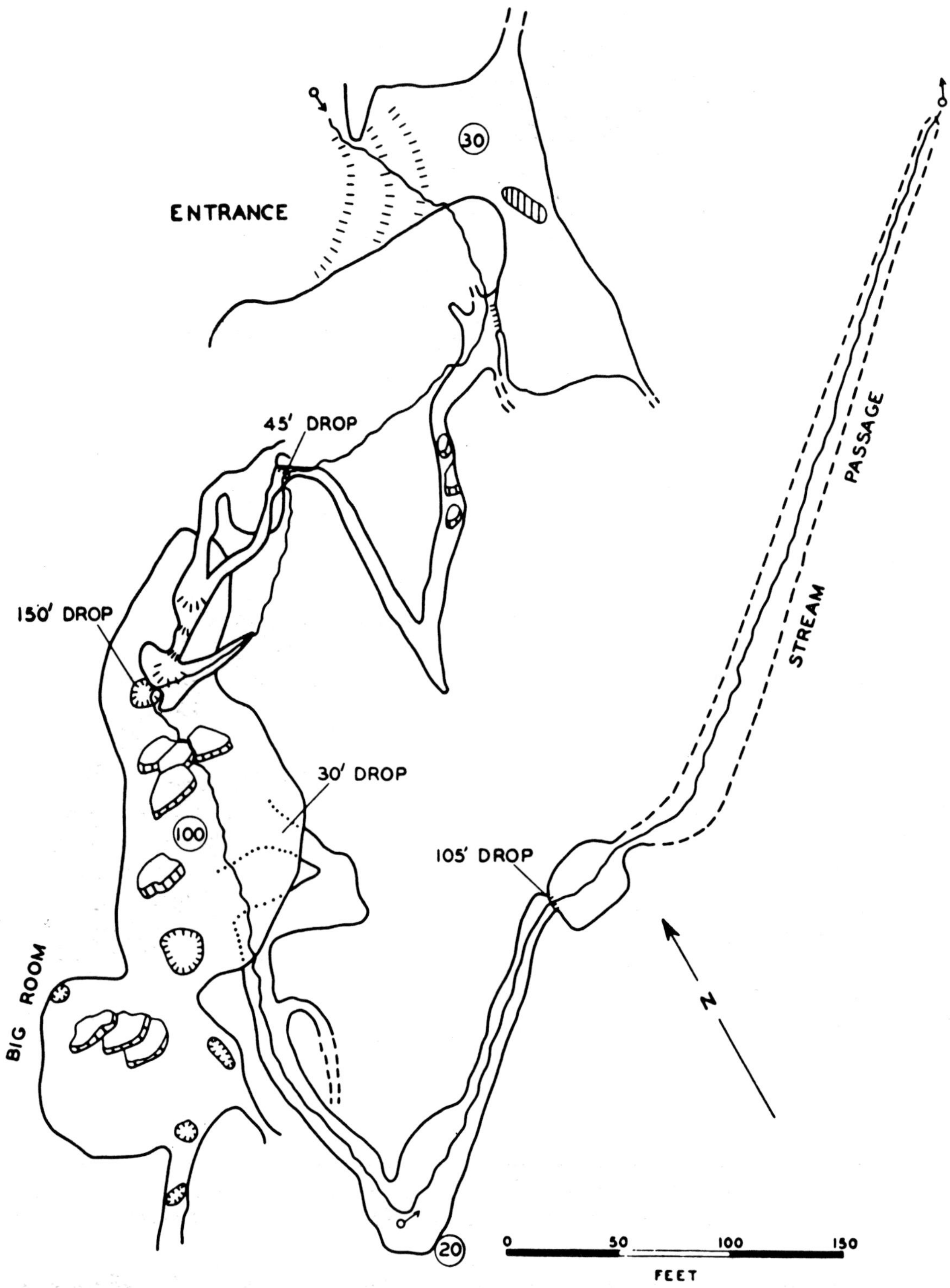

Figure 9.16. 1955 map of Bull Cave from Barr's Caves of Tennessee (1961), *page 76. Survey by J.H. Gibbons and T.C. McDaniels.*

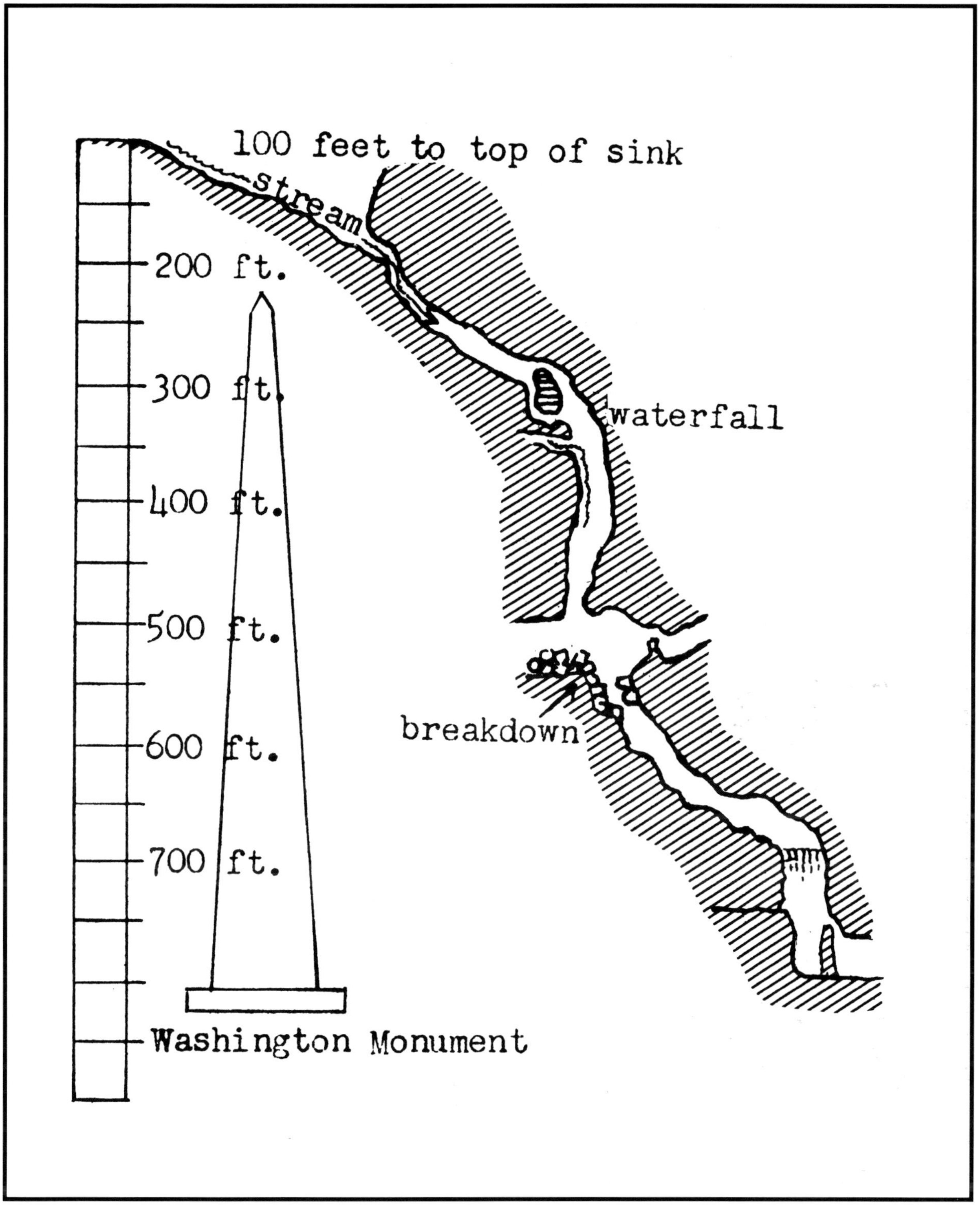

Figure 9.17. 1955 profile of Bull Cave. From "Down Into Bull Cave," by John H. Gibbons and Daniel Hale, NSS News, *Vol 14, No.6 (1956) p 62.*

tains National Park, it is closed to the public and permits to enter the cave are difficult, if not impossible, to obtain. Bull Cave is an exceptionally dangerous cave, due to the many deep drops in the cave and the cold, wet conditions that are encountered.

A Description of Whiteoak Blowhole

J.D. Lawrence Jr wrote the description of the Blowhole in Barr's *Caves of Tennessee* (1961). This is another cave that was explored by the National Speleological Society members on their expedition to the Great Smoky Moun-

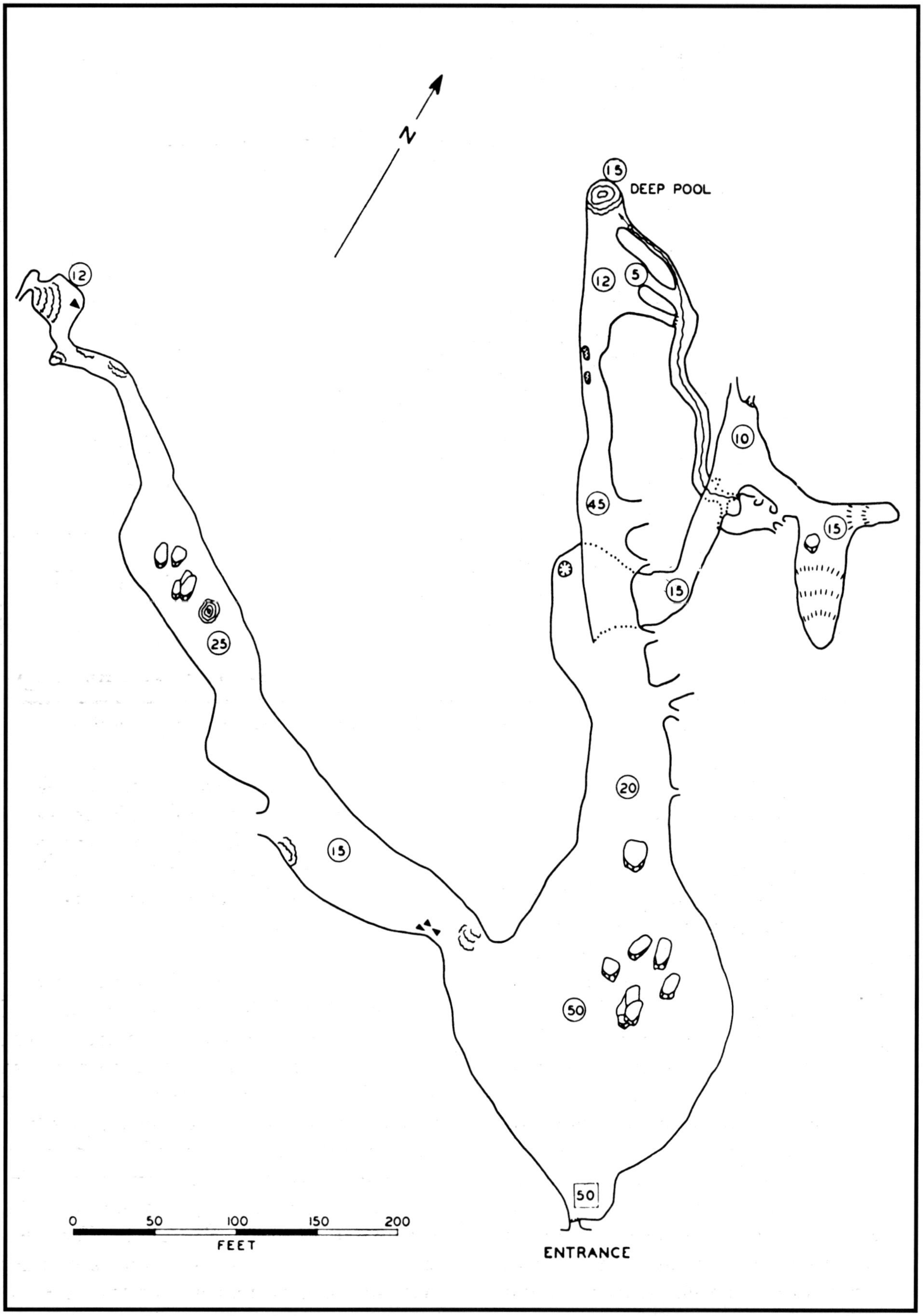

Figure 9.18. 1950 map of Whiteoak Blowhole from Barr's Caves of Tennessee *(1961), page 75. Survey by E. Bleichen, R. Rowland, and J. Lawrence.*

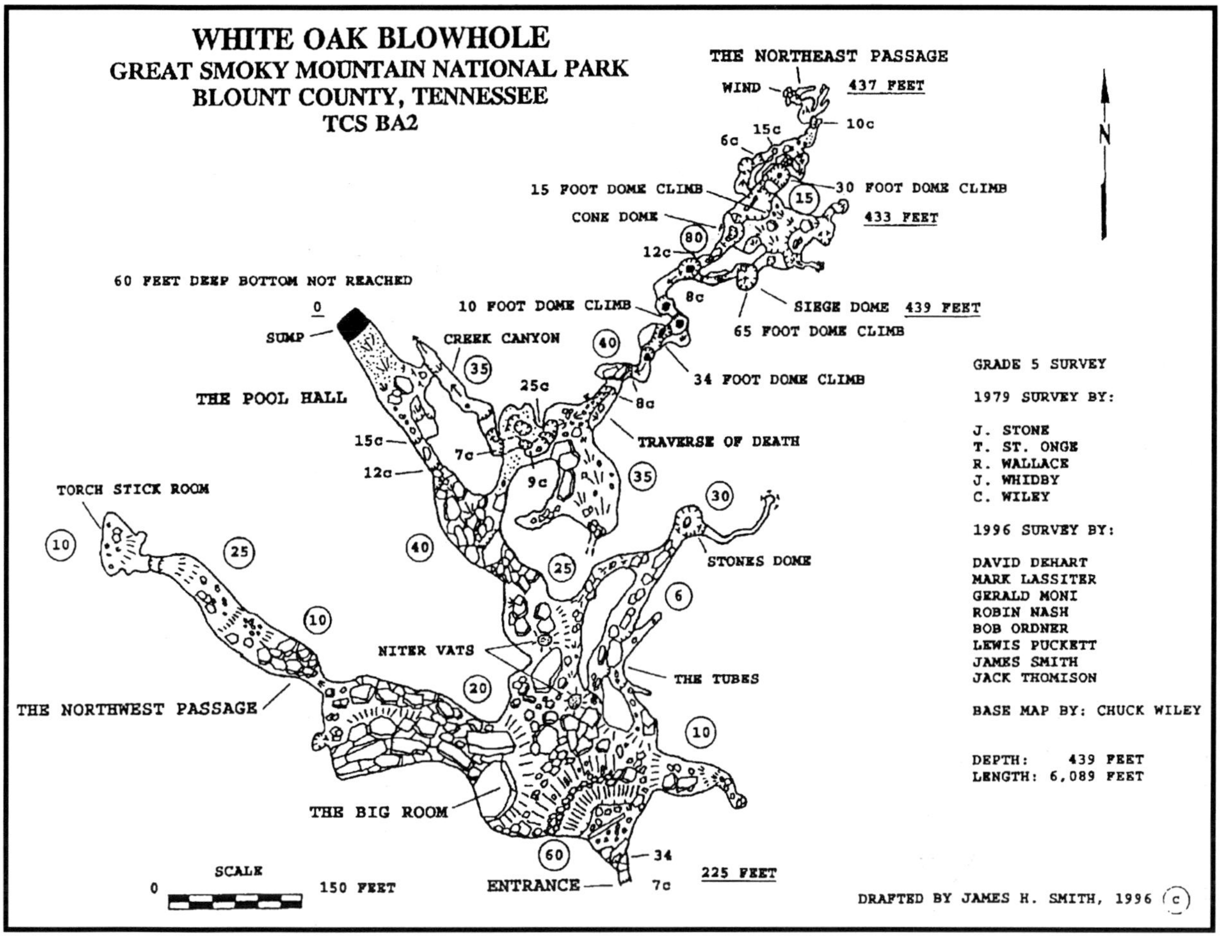

Figure 9.19. 1996 map of Whiteoak Blowhole. © 1996 James H. Smith, used by permission.

tains National Park in 1950. Lawrence is listed as one of the twelve Society members on that December 2–5, 1950 trip. The map in *Caves of Tennessee* is dated 1950 and is credited to E. Bleichen, R. Rowland, and J. Lawrence. Bleichen is also identified in the *NSS News* article as being present on the trip. The description of this cave by Lawrence is given below:

> The entrance is a rectangular hole which drops vertically for 50 feet into a room 30 feet high, 160 feet wide, and 200 feet long. The main gallery extends northwest for 615 feet and terminates at a large pool. The pool is fed by a small stream which issues from a narrow, southeast side passage. The stream passage is connected with another side passage which leaves the main gallery about 350 feet from the mouth. Several crawlways and high leads open into this section of the cave.
>
> To the left of the entrance another gallery runs west-northwest for 730 feet to a dripstone grotto. This gallery has an irregular boulder floor, with pools and a small stream, and is decorated with some flowstone and stalactites.[12]

The 1950 map for the Blowhole is shown in Figure 9.18. This cave has been renamed Whiteoak Blowhole to avoid confusion with a dozen other caves in Tennessee that are also named Blowhole Cave. More recent exploration has shown this cave to be 1.2 miles long and 439 feet deep. It is an Indiana bat hibernaculum during the winter months. James H. Smith and others resurveyed the cave in 1996. This map is shown as Figure 9.19. The 1996 map shows much more detail than the earlier map, including the remains of saltpeter leaching vats. Apparently, the original local name

12 Barr, op. cit., pp 74–75, 77.

Figure 9.20. Erica Sughrue by large flowstone formations in Gregorys Cave. Photo by Bob Biddix, May 17, 2008.

for this cave was Saltpeter Cave. Entrance to Whiteoak Blowhole is prohibited without a permit from the National Park Service.

Bats of the Great Smoky Mountains

Alan Rabinowitz and Ben Nottingham conducted a study of bats in the Great Smoky Mountains in 1978.[13] As a result of that study, we know which caves are being used by what species of bats.

Bull Cave is used by the eastern pipistrelle bat (*Pipistrellus subflavus*) and the Indiana bat (*Myotis sodalis*).

Gregorys Cave was used by eastern big-eared bats (*Plecotus refinesquii*) in 1978, but they have not been observed there since. The eastern pipistrelle bat (*Pipistrellus subflavus*) is regularly seen in the cave.

Whiteoak Blowhole is used by the eastern pipistrelle bat (*Pipistrellus subflavus*) and the Indiana bat (*Myotis sodalis*). The eastern pipistrelle bat population was estimated at between 4,000 and 5,000 individuals. The Indiana bat population was estimated at approximately 12,000 to 15,000 individuals and is one of the ten largest known hibernating colonies of Indiana bats in the United States.

Entering a cave while bats are hibernating (generally from September 1 through April 1) may disturb the bats and cause them to awaken out of hibernation. This uses up the fat supply that they depend upon to survive through the winter and they may starve to death. Please do not enter any bat caves during this time of the year. In addition to that restriction, all caves in the Great Smoky Mountains National Park require a permit to enter.

13 Alan Rabinowitz and Ben Nottingham, "Human Visitation and Fall/Winter Cave Usage by Bats in the Great Smoky Mts. National Park," *Speleotype*, vol 13 (1980), no. 1 & 2, pp 4–20.

Directions

From the Sugarlands Visitor Center, in Great Smoky Mountains National Park, drive west on the Little River road for 17.9 miles to an intersection. The road to the right leaves the Park and leads into Townsend. The road to the left, Laurel Creek Road, continues on to Cades Cove. After 7.6 miles you will enter Cades Cove. The road to the left leads to the Cades Cove Campground, a picnic area, and the Cades Cove Ranger Station. The Cades Cove Loop Road, that continues straight ahead, is a one-way loop that circles the edge of Cades Cove. After 2.1 miles, you will see a gated dirt road to the right. Park here, but do not block the gate or the road. From here, you must hike to reach Gregorys Cave.

Figure 9.21. Erica Sughrue by beautiful flowstone formations in Gregorys Cave. Photo by Bob Biddix, May 17, 2008.

After 0.1 mile, you will encounter the foundation of an old house. Continue 0.3 mile to the base of the mountain. Here you will encounter limestone rock outcrops and the entrance to Gregorys Cave. Gregorys Cave is locked and gated. Entrance to Gregorys Cave is prohibited without a permit from the National Park Service.

You may also easily reach Gregorys Cave by entering Great Smoky Mountains National Park from Townsend on State Highway 73 West. 0.1 mile after entering the Park, turn right at the "T" intersection and follow Laurel Creek Road to Cades Cove.

Side Trips

Cades Cove, where Gregorys Cave is located, is one of the **must visit** sites in Great Smoky Mountains National Park. Cades Cove is a flat-bottomed cove that is 5 miles long and 2 miles wide. It was cleared by the early settlers and converted into farmland. Currently, the Park Service is maintaining the farmland as pasture, which preserves the breathtaking views of the surrounding mountains. A loop road goes around the edge of the Cove. The Cove

Figure 9.22. A portion of the Main Passage in Gregorys Cave. Erica Sughrue is in the foreground and Larry Matthews is in the background. Photo by Bob Biddix, May 17, 2008.

has dozens of pioneer homes, barns, mills, and other structures that have been maintained to show visitors how people lived in this area over 100 years ago. You can watch corn being ground into corn meal at the John P. Cable Mill and even buy a bag of freshly ground corn meal to take home with you. Unlike the finely ground corn meal sold in stores, this meal is ground on real millstones and is much more coarse. There is a wonderful, traditional recipe for cornbread on the side of the bag. Try it. You will like it.

Cades Cove is also one of the best areas in the park to view wildlife. Deer are abundant and frequently seen here. Wild turkeys and black bear are also seen here on a regular basis.

If you have time, the Rich Mountain Gap Road turns off from the main Cades Cove road near the far end of the Cove. The turnoff for this road is well marked. This gravel road climbs steeply out of Cades Cove and offers spectacular views back into Cades Cove and south into the heart of the Great Smoky Mountains. After approximately 6 miles the road peaks at 1,940 feet elevation at Rich Mountain Gap. This marks the boundary of the Park. As you continue down the road on the far side of Rich Mountain Gap you will go down a winding mountain road into Tuckaleechee Cove.

The most popular trail in Cades Cove follows Abrams Creek downstream for 2.6 miles to Abrams Falls. Abrams Falls is easily one of the most beautiful waterfalls in the Smoky Mountains and its plunge pool makes a delightful swimming hole in the warm summer months. Pack a picnic lunch and spend several hours in this wonderful place.

Figure 9.23. Erica Sughrue by beautiful flowstone formations in Gregorys Cave. Photo by Bob Biddix, May 17, 2008.

Figure 9.24. Erica Sughrue by a small pool and formations in Gregorys Cave. Photo by Bob Biddix, May 17, 2008.

Chapter 10

The Lost Sea

The Discovery of Craighead Caverns

Craighead Caverns, as The Lost Sea was once known, was named after a Cherokee Indian Chief, named Craighead, who once owned the property on which the cave is located. Chief Craighead acquired the land through either the Ocoee or Hiwassee Land Grants. The cave had a small natural opening on the side of the mountain that opened into a series of very large rooms. Craighead may have discovered this tiny entrance, but there is no historic record of who first explored the cave. Artifacts found in the vicinity of the "Council Room" around 1927 or 1928 included pottery, arrowheads, jewelry, and other artifacts left by the Indians. Craighead Caverns is located 25 miles from "Chotow," the historic capital of the Cherokee nation.

When the first white settlers arrived in this area in the 1820s, they also discovered the cave and used the "Sand Room" as a storage place to store potatoes and other vegetables. The constant 58-degree temperature and absolute darkness provided a natural refrigeration system for the food.

Saltpeter Mining During the Civil War

During the Civil War, Craighead Caverns was mined for saltpeter. Saltpeter is a naturally occurring mineral, $Ca(NO_3)_2$, that accumulates in the soil of some caves. Since it is highly soluble in water, this mineral accumulates in significant quantities only in the drier portions of caves. Saltpeter was mined from caves in Kentucky, Tennessee, Virginia, and West Virginia as early as the War of 1812. Records for this time period are poor, so it is possible that Craighead Caverns was also mined during the War of 1812.

In order to mine saltpeter from Craighead Caverns, it would have been necessary for the miners to construct one or more leaching vats. These would have been constructed from local wood using simple hand tools. Over a hundred Tennessee caves are known to have been mined for saltpeter and the leaching vats are still present in some of those caves. Therefore, we know that there were two commonly used types of saltpeter vats: square and V-shaped. Photographs of a variety of these vats can be found in *Descriptions of Tennessee Caves.*[1] At some caves, the leaching vats were located outside or at the entrance. There are no original leaching vats remaining inside The Lost Sea. However, a reproduction of such a leaching vat is on display in the Keel Room.

Once the leaching vats had been built, the miners would excavate saltpeter-bearing dirt and place it into the vat until the vat was nearly full. Normally, the vat was lined with straw to help prevent dirt and water from leaking through the cracks between the wooden planks. Water was then slowly poured over the dirt in the vats. The water would dissolve the highly soluble saltpeter as it worked its way downward. This saltpeter-enriched water, called "liquor," would then be collected in wooden water troughs at the base of each leaching vat.

The next process was to pour this liquor through another, smaller leaching vat which was full of hardwood ashes. These ashes are rich in potassium and this process would convert

---

1 Larry E. Matthews, *Descriptions of Tennessee Caves*, Bulletin 69, Tennessee Division of Geology, 1971.

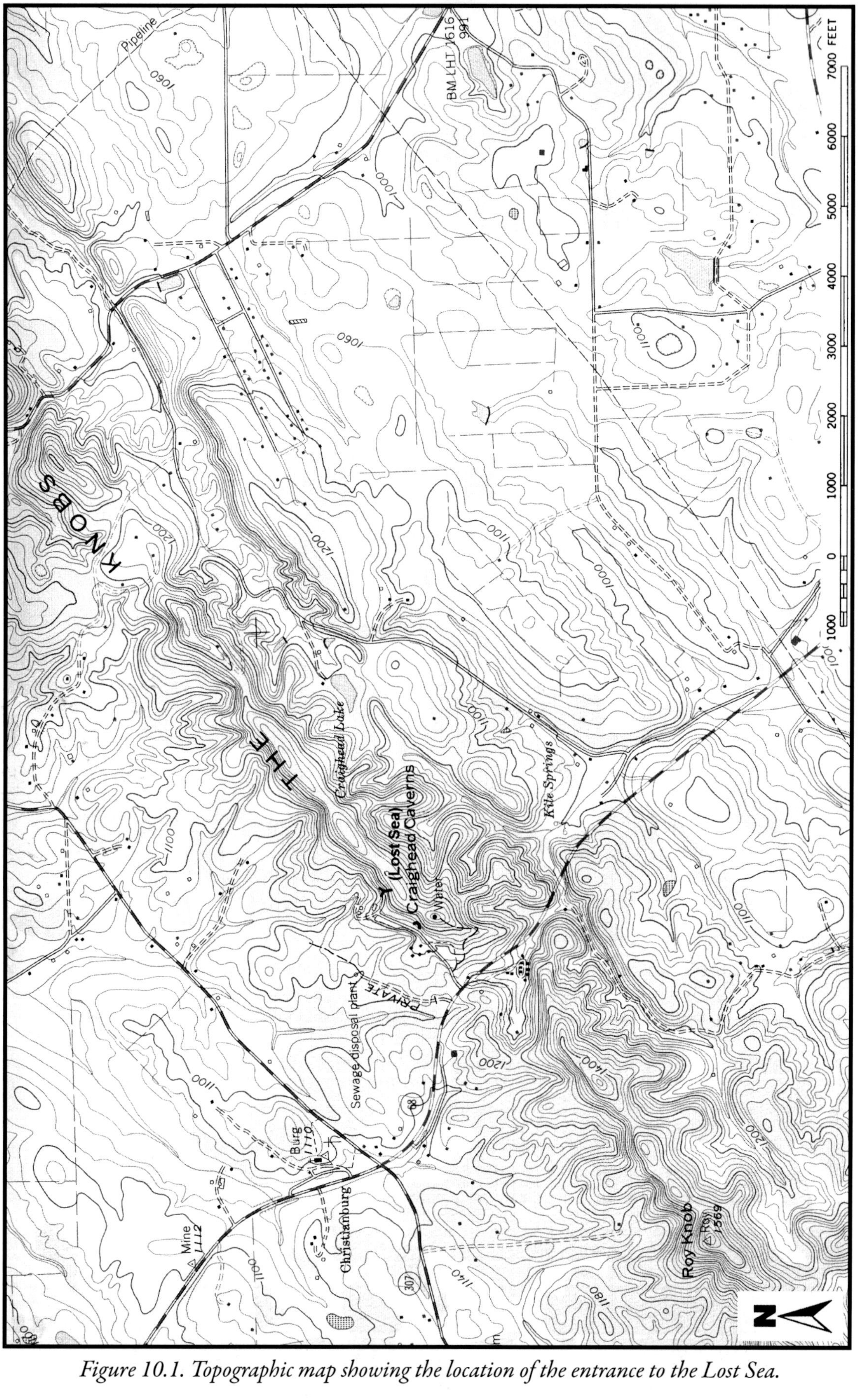

Figure 10.1. Topographic map showing the location of the entrance to the Lost Sea.

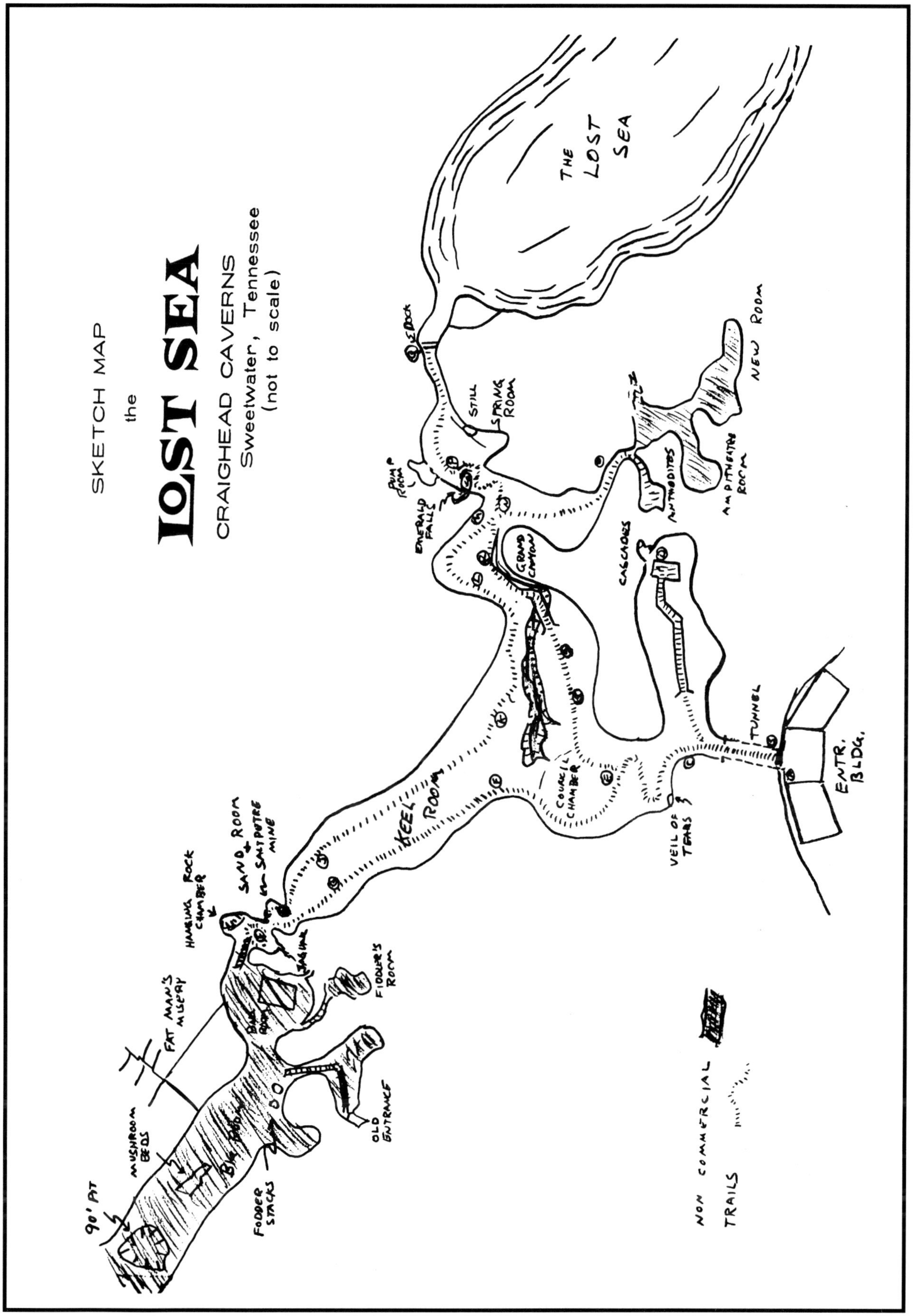

Figure 10.2. 1929 map of Craighead Caverns by Berlen C. Moneymaker with 1940 additions by W. E. Michael.

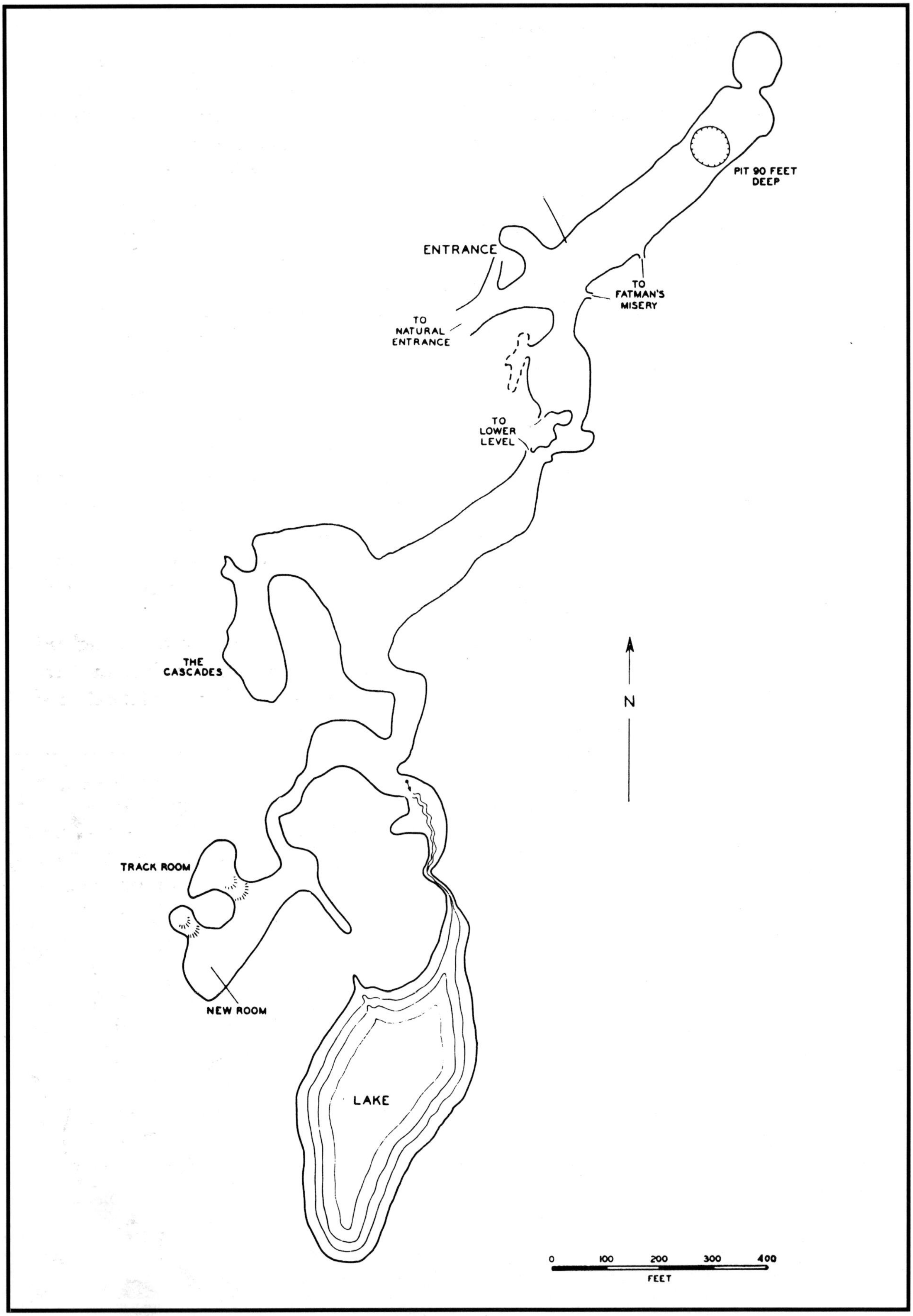

Figure 10.3. 1955 map of Craighead Caverns by Guy Crawford.

the cave saltpeter, $Ca(NO_3)_2$ into potassium nitrate, K_2NO_3, which is the type of saltpeter necessary for the manufacture of gunpowder. This process also caused some other unwanted minerals in the liquor to precipitate out, so that they could be easily removed by straining.

The resulting liquid was then placed in large iron kettles, where it was carefully boiled until much of the water evaporated, leaving behind crystals of saltpeter. These crystals were removed, dried, and packed in bags, so that they could be shipped to a powder mill. At the powder mill, the saltpeter crystals were ground with charcoal and sulphur to make gunpowder.

One of the more intriguing stories of the cave's use during the Civil War is that of a Union spy who penetrated the guarded cave and nearly succeeded in blowing up the mining operation before he was captured. According to the story, which was recorded in a diary, he was shot near a large gum tree at the cave entrance.

The Lost Sea, in their booklet *The History of the Lost Sea Adventure*, provides the following information about the above mentioned diary and saltpeter mining in the cave:

Over one hundred years ago, instead of tourists dressed in gaily colored sports clothing coming from the entrance of Craighead Caverns near Madisonville and Sweetwater, Confederate men came out of the entrance to dodge enemy guns with 50-pound cans of gunpowder on their backs.

The story is contained in an old diary which was found previous to 1934 by the late George Coltharp. It had been kept by his father, the Rev J.H. Coltharp, who was in charge of manufacturing gunpowder in the Caverns, now The Lost Sea.

Yellowed with age, the old diary had been used as a scrap book. The members of the Coltharp family had no idea that the interesting stories were under the clippings until a member accidentally tore one of them away. The owner of the diary is still unknown. Perhaps someone who reads this story can provide The Lost Sea with this information.

It has long been said that Craighead Caverns is the only place in this area where gunpowder was made up until the close of the Civil War.

Figure 10.4. The date "1863" smoked on the ceiling in the area mined for saltpeter. Photo by Larry E. Matthews, March 12, 2008.

Figure 10.5. Old names smoked on the ceiling in the area mined for saltpeter. Photo by Larry E. Matthews, March 12, 2008.

Distributing the gunpowder, the most dangerous work, was the most profitable. The powder was distributed to the Confederate army near Knoxville. The powder was put in cans with a screwed-on top, the cans weighing about 50 pounds. The men carried the heavy cans of gunpowder on their backs, dodging the Union men, and traveling only by night. Three days and nights were required to make the less than 50-mile trip to Knoxville according to the record. The men received $9 per night. Among the men who distributed gunpowder were Asberry Kile, Russell Hicks (grandfather of Amos Wilson), G.H. Coltharp, and Wilse Lane.

These men, in taking the gunpowder to Knoxville, went by way of the Morgantown ferry on the Little Tennessee River. The ferryman, afraid to set them across the river, would leave the boat and the men would swim across and bring back the boat.

A large amount of gunpowder was also distributed to the Confederate army stationed near Chattanooga. Men from the cave took gunpowder to all parts of the South. J.P. Guffey, veteran of the Civil War, stated that he saw men from Craighead Cave carrying gunpowder near Atlanta.

According to the record, the men in the cave worked by contract, except the Rev Coltharp, who received $25 per month for overseeing the making of the powder. Saltpeter, used in making the gunpowder, was dug from the walls of the cave near a large underground lake and carried a mile to the entrance of the cave. Getting out 14 bags of the dirt, for which the men were paid 75 cents, was considered a day's work. They were paid extra for all they got over 14 bags. The bags are said to have held about a bushel.

The men worked frantically when there was special need for more gunpowder, according to the old diary. When it was necessary for them to put in more time their provisions were brought to them. They often worked for days without sleep.

Asberry Kile, who is mentioned frequently in the diary, had a mule on which he took provisions to the cave. One day, while bringing food to the cave, the

Federals began to fire on him and he left his mule and dashed into the cave. The mule, possibly taken by the Federals, was gone five days before it returned.

After the dirt containing the saltpeter was taken from the cave, it was put into a wooden chute 14 inches wide and 6 inches high and run about 30 yards down the knob into the valley where it was made into gunpowder. In the valley below the cave may be seen the effects of the saltpeter where the ground has been "killed."

It is said that the dirt was placed in a wooden hopper much like the old fashioned hoppers used in obtaining lye for making soap. Water was poured on the dirt and the "lye" was caught to be boiled down until all the water was taken out, leaving crystals. These crystals were mixed with charcoal and ground together making gunpowder.

Large quantities of ashes were used in making gunpowder. The men paid 30 cents per bushel for the ashes. The record shows 13 bushels purchased at one time from William McKeehan.

According to the diary, G.H. Coltharp and Wilse Lane were hemmed in the cave without food. The knobs around the cave were guarded by Confederates, but sometimes the Federals could make their way around the guards. However, because no Yankee dared go into the cave, they knew they would be safe. Anyone who has seen the cave cannot doubt the safety they must have felt; the numerous passageways leading off from the main rooms furnished ideal places for hiding. They knew they could be found only by someone familiar with the cave.

Frequent efforts were made to disturb the work in the cave. One time a Union spy made his way by the guards and had dynamite placed, ready to blow up the cave when he was discovered. He was tied to a large gum tree near the cave and shot.

Names which were given in the diary represented a large number of Monroe County families. A.L. McKeehan, who dug saltpeter from the cave, was the grandfather of T.A. McKeehan, C.W. Hicks worked in the cave. He was the father of Judge Sue K. Hicks and Mrs Margaret Haggard of Madisonville. J. Franklin Magil, who also worked in the cave, was the father of the late W.W. Magil.

It is interesting to note the contrast in prices then and now. Domestic used in straining the "lye" cost 66¾ cents per yard. Beeswax, said to have been used to line the sacks used in carrying the dirt from the cave was quite expensive. The record shows the purchase of an axe was $4.

There was quite a difference in the price of groceries then and now. The men boarded with a Mrs Mercer, who lived near the cave. The board for all seems to have been taken from the funds received from the gunpowder. The record shows the purchase of 117¼ pounds of beef at one time at 8 cents per pound and 32 pounds of pigs' feet at 4 cents per pound. Bacon was 7½ cents per pound. Salt brought 55 cents per pound and could not be found except as brought in. The dirt from the floor of the smokehouses was dug and boiled to get the salt. Small sweet potatoes were sliced and parched and used for coffee. Board was then $8.66 per month.

This is truly a remarkable account of the saltpeter mining activities at Craighead Caverns. Very few other caves have so many details surviving about their saltpeter mining operations.

The Discovery of the Lost Sea

Throughout the early history of the cave there were rumors of a large underground lake somewhere deep within the cave. This rumor was confirmed in 1905 when a 13-year-old boy name Ben Sands crawled through a 300-foot-long crawlway and emerged in a huge room half filled with water. The room was so large that his light could not reach the far walls. According to legend, for the rest of his life Sands delighted in describing how he threw mud balls as far as he could into the blackness and heard nothing but splashes in every direction. Ben Sands' father, Johns Sands, had explored the cave before him, and occasionally acted as a guide for persons visiting the cave.

We know now that the water level in this huge lake naturally fluctuates up and down from season to season and from year to year. Therefore, the crawlway that leads to the lake was sometimes flooded and not passable. Modern development of the cave required drilling a tunnel to drain excess water, so that the lake could be reached at all seasons of the year.

In 1915 Mr George Kyle became owner of the property and was the first person to attempt to develop the Craighead Caverns for commercial purposes. He installed a wooden dance floor in the "Big Room," which is a large room near the natural entrance. Craighead Caverns was one of several caves developed as a dance cave in Tennessee. A steep trail with over 100 steps leads down from this entrance into the main level of the cave. In the days before air conditioning, a night dancing in Craighead Caverns in the 58-degree cave air was a rare treat. Cockfights are reported to have been another frequent activity in the cave. At the same time, there are reports that moonshiners were operating stills inside the cave.

In 1927 Craighead Caverns was opened as a true commercial cave. A larger and more accessible entrance was opened below the natural entrance to make entering the cave less strenuous. The steep trail from the natural entrance, with over 100 steps, no longer had to be used. The cave was electrically lighted by the Tennessee Power Company at this time. The success of this business was cut short by the Great Depression, and Craighead Caverns closed in 1930.

The next commercial venture in the cave was a mushroom farm which was operated in the Big Room from 1939 to 1940. Horse manure was supplied from Fort Oglethorpe, where many horses were stabled, to provide fertilizer for the mushrooms. When Fort Oglethorpe became mechanized, fertilizer was no longer available and the mushroom farm was closed. All mushroom beds were removed except for one, when the second dance floor was built in 1947. This new operation was known as the "Cavern Tavern."

In 1940 to 1955, Van Michael, who was then just a boy, began to dream of developing the cave. He operated the cave for a couple of seasons and took a thousand or so visitors through the cave. In September of 1964, the Craighead Caverns Company was formed and development began in December. Much of this development was done by Roy A. Davis, who had opened Cumberland Caverns in middle Tennessee in 1957 and was an expert in cave development. The Lost Sea opened as a tourist attraction in June of 1965.

During the height of the Cold War, the Federal Government not only considered caves for use as fallout shelters, they actually went so far to designate a number of caves as official public fallout shelters. A few of these caves, including The Lost Sea, were stocked with supplies. The theory was that if people had enough warning, they could go to one of these designated caves and be protected from the blast of an atomic bomb when it exploded and then spend several more weeks protected from the radioactive fallout.

Not only was The Lost Sea designated as a nuclear fallout shelter, it was actually stocked with supplies. At one point in the cave, there is still a rather large stack of tins filled with fortified crackers, intended to provide food during the stay in the cave. These stockpiles also normally included 17.5-gallon water cans, medical supplies, and radiation monitoring devices. However, at the present time, the tinned crackers are all that remain in the cave from its fallout shelter days in the 1960s. According to cave manager Lisa McClung, The Lost Sea was rated as capable of holding 20,000 people in case of nuclear attack.[2]

A Jaguar in the Cave

In 1939, the bones and footprints of a giant Pleistocene jaguar were found deep inside Craighead Caverns. George Gaylord Simpson described this discovery in 1941:

2 Lisa McClung, Personal communication, March 12, 2008.

Figure 10.6. Tins of crackers left from the fallout shelter stockpile in the Lost Sea. Photo by Larry E. Matthews, March 12, 2008.

The remarkable discoveries here recorded were made in the Craighead Caverns, about 5½ miles southeast of the town of Sweetwater in Monroe County, eastern Tennessee. This cave, much the largest in this region, has been known for many years. Indian relics have been found near the entrance and the Indians certainly knew the cave and used it at least occasionally for shelter, but there is no evidence that they explored it extensively. The name Craighead is said to be that of a former owner, a Cherokee Indian. Saltpeter was mined here, as in many southern caves, to make gunpowder during the Civil War. The main rooms have been open to the public for some 50 years or more and the cave enjoys increasing fame as a tourist attraction. Cock fights and dances were formerly held in one of the rooms near the entrance, now used for mushroom growing. In spite of all this, parts of the cave were discovered only recently and some parts are still unexplored.

The discoveries to be discussed in this paper were made in 1939 and 1940 by two young men employed at the caverns, Jack Kyker and Clarence Hicks. Living in the vicinity and inveterate cave-crawlers, familiar with this cave from boyhood, they devoted their spare time to searching for new rooms and in the course of this pursuit they found some bones in a narrow fissure of extremely difficult access. In another part of the cave, almost equally inaccessible at that time, they found footprints of some large animal.

These discoveries were brought to the attention of the present owners of the cave, Dr W.J. Cameron and Mr W.E. Michael of Sweetwater, who recognized their importance, investigated the occurrence,

Figure 10.7. George Gaylord Simpson at his American Museum desk in the early 1950s.

discoveries and the jaguar localities. Dr Raymond Ditmars and his assistants at the New York Zoological Society permitted and assisted the obtaining of tracks of a living jaguar for comparison.

and removed some of the bones. They brought the latter to the American Museum[3] for identification and Dr Walter Granger and I found that the bones included parts of a very large jaguar and of an elk (or wapiti) fawn. In May 1940, I visited Sweetwater and through the hospitality and cooperation of Dr Cameron and Mr Michael I was able to examine the sites and to obtain further remains of the big jaguar as well as a plaster mold of the footprints. These specimens, as well as those first found and some recovered at later dates, were generously presented to the Museum by Dr Cameron and Mr Michael.

I am also indebted to Dr Berlen C. Moneymaker for a copy of his plane table map of Craighead Caverns as known in 1929, here published with additions by Mr Michael to show some of the later

Occurrence of Bones

The Craighead Caverns are a large solution cave, in one of a series of knobs formed by outcrops of Paleozoic limestone. As shown in the accompanying map, the cave consists of an irregular series of large rooms, the lowest of which extends below groundwater level and contains a navigable underground lake. (Through fluctuations in water level the entrance to the lake room was formerly completely under water and its discovery is recent.) The outside entrance to the cave is about 1,100 to 1,200 feet above sea level. The passages between the rooms are narrow in some cases and in others are low and nearly or quite filled by fine red silt, so that some of the rooms remained unknown or unexplored until recently. There are also numerous fissures through which a man can barely or cannot squeeze. One of these, barely wide enough for a man and for the most part so low as to require crawling on face and abdomen, leaves the main cave at the point marked "C" on the map and winds for perhaps 300 feet to a point below "B" on the map, following approximately the dotted line. The passage here communicates with a

3 He is referring to the American Museum of Natural History in New York City.

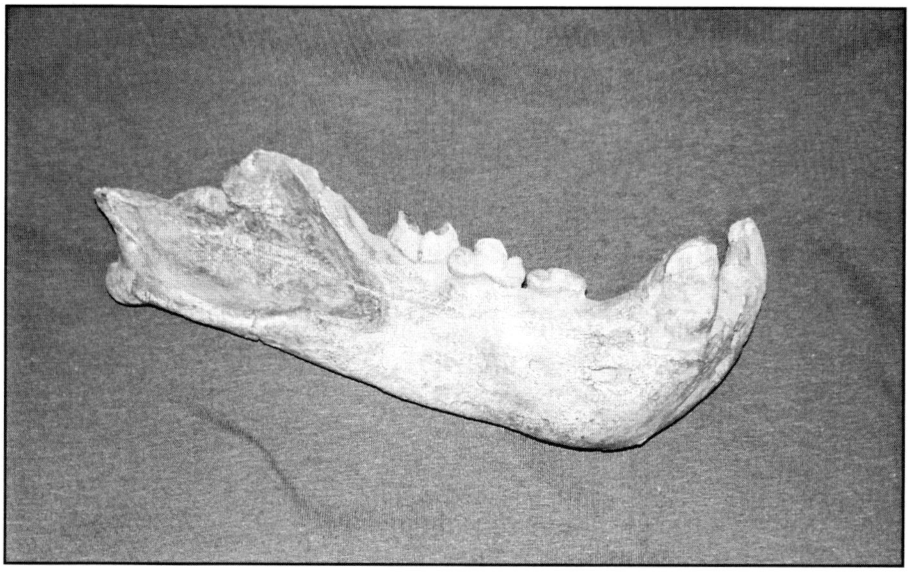

Figure 10.8. A cast of the jaguar jaw found in Craighead Caverns. Photo by Larry E. Matthews, March 12, 2008.

vertical fissure (not shown on the map) of considerable height and depth but still very narrow, developed more or less along the northeast-southwest joint system that can be detected from the orientation of several of the cave's rooms and fissures. The small room shown at and below and to the left of "B" on the map is above this fissure, with which it communicates by a small hole in the floor.

The bottom of this fissure is filled with red cave earth and the first bones found were partly buried in and partly protruding from this earth at the lowest point in the floor of the fissure. These include a jaguar lower jaw and some 25 or 30 fragments, apparently all of a single animal, a very young, probably newborn, elk (or wapiti) fawn. This point is approximately 225 feet lower than the present entrance to the cave and about 250 feet below the old entrance shown on the map, now closed by fallen rock. It is only about 200 feet horizontally from the present entrance and about 250 from the old entrance, although by the shortest known passages it is at least 400 feet from the former and 650 from the latter.

To the left of this point (by the orientation of the map) the floor of the fissure rises rapidly and it becomes small and irregular, finally so small that we could find no further way by which a man or a jaguar could possibly continue along it. The impracticable opening above this probably communicates with the sloping room leading to the old entrance to the left of the new entrance, although we could not certainly prove this. At the highest accessible point, about 50 feet above the first discovery, in a deposit of very wet red clay, part of the skull of a jaguar was found.

From these relationships, it would be just possible that the remains were washed down from the passage leading to the old entrance, or even from outside the cave through that entrance, but this seems unlikely. It is possible that the jaguar or jaguars wandered alive into the small room above the fissure, fell into the latter through the hole in the floor of the room, and were unable to get out. In this case the individual represented by the skull must have been trying to climb out and became wedged or died of starvation at the highest point that it could reach. The fawn may similarly have wandered in and fell or, more likely, may have been dragged into the cave by a jaguar. Its bones were broken when found and may have been broken by a jaguar, but this is uncertain. Some of the jaguar bones were gnawed by rodents after death.

On the wall of the fissure, within reach of the bottom, there are several parallel grooves, now lightly coated with travertine. The discoverers believed that these were claw marks made by the jaguar in its endeavors to escape, a distinct possibility but one that I see no way of proving or disproving.

Although there is no duplicate of parts and both are surely of the same kind of animal, the remains may represent two individual jaguars, since the skull appears to be of a somewhat younger individual that the lower jaw and not to articulate well with the latter. These bones will be figured and described in a following paper in which the occurrences and classification of all the larger Pleistocene felines of North America will be discussed. The elk bones evidently represent a single individual and do not appear to differ significantly from recent *Cervus canadensis*.

Occurrence of Footprints

Besides the scratches on the wall near the bones, there are numerous other scratches in various parts of the cave. Some of these appear to have been made by claws and it is possible that a few of the larger ones were made by the jaguar, but this is not subject to proof. Such single scratch markings in the parts of the cave long explored and open to the public must in any case be subject to doubt. Two clear footprints were, however, found in such a position and of such a nature that they surely represent an animal and probably the jaguar. These occurred in a different part of the cave from the bones, at the point marked "A" on the map, some 800 feet distant in a straight line horizontally and about 1,500 feet by a route practicable for a man or a jaguar. The distance by a practicable route from the present entrance to the cave is about 1,600 feet and from the old entrance about 1,800 feet. It is possible that a shorter route existed when the footprints were made but this is unlikely. Intensive exploration has revealed no trace of a shorter passage to the surface and the tracks are on a pile of fallen rock and cave clay suggesting that no extensive remodeling of the topography has occurred in this part of the cave since they were made.

The footprints are in a fissure opening into what is called the New Room. This was known before, but was difficult of access and seldom entered until 1940 when a trail was dug through the thick clay deposit that almost sealed the corridor leading into the room. The main floor and ceiling of the room are roughly horizontal but on one side the ceiling slopes upward rapidly and the floor of this fissure-like extension is also steeply sloping and formed by fallen blocks of limestone between and sometimes covering which is much red cave earth and clay. The fissure narrows at the top and finally is nearly closed by rock and by clay evidently washed down from above. It here becomes impassable for a man or any animal as large as a jaguar. Aside from vague markings that may or may not be footprints, two clear, partly superposed prints were found just below this highest accessible point. The prints are those of an animal traveling straight up the slope. Before reaching this stretch of softer material the animal would have been traveling on rock or on harder earth where the prints would be less distinct. The elevation above the floor of the New Room is about 30 feet. The suggestion is that the animal was lost and climbed the slope in an endeavor, here vain, to find some way out. Darkness here is, of course, absolute and there is no place in the cave more than a few feet beyond the entrance where any glimmer of light penetrates.

The material in which the prints were made is not a pure, slick clay but a sort of earth, clay mixed with very fine sand, of the brick-red color of all the cave earth and most of the soil of the countryside overlying the limestone. The earth is still soft and moist and can have hardened little if any since the prints were made. They were clear and sharp and looked as if they might have been made only a few minutes before their discovery, except that a small amount of loose, almost dry earth and darker dust had drifted into them. In view of the complete absence of dust in the air, even this trivial accumulation might suggest considerable age or a marked change in conditions since the prints were made, but it seems very improbable that there has been any radical change. The presence of some dust in the tracks is consistent with relatively great age, but does not necessarily indicate this. Within historic times even these remote parts of the cave have been frequented both by rats and by bats and their activities could produce such an effect relatively rapidly.

Since prints in such soft earth were sure to be damaged or destroyed soon, now that the New Room is being opened to tourists, their permanent preservation was assured by filling them carefully with plaster and removing this along with such loose dirt as adhered to it. This mold was carefully cleaned by Otto Falkenbach in our laboratory, a rubber cast made and reversed, and plaster replicas made that faithfully reproduce every detail of the originals.[4]

Figure 10.9. Lisa McClung in the Entrance Tunnel to The Lost Sea. Photo by Larry E. Matthews. March 12, 2008.

The jaguar bones found in The Lost Sea are estimated to be 20,000 years old. Surprisingly, jaguar skeletons have been found in ten other Tennessee caves: Baker Bluff Cave, Sullivan County; Big Bone Cave, Van Buren County; Johnson Cave, Putnam County; Little Airplane Cave, Hamilton County; Post Office Cave, Maury County; Ruby Falls Cave, Hamilton County; Salt River Cave, Franklin County; Sinking Cove Cave, Franklin County; Tumbling Shoals Cave, Hamilton County; and Wolf River Cave, Fentress County. Clearly, jaguars must have been abundant in Tennessee at some time in the past.

4 George Gaylord Simpson, "Discovery of Jaguar Bones and Footprints in a Cave in Tennessee," *American Museum Novitates*, No. 1131 (August 6, 1941), pp 1–12.

Figure 10.10. Plaque designating The Lost Sea a Registered Natural Landmark. Photo by Larry E. Matthews, March 12, 2008.

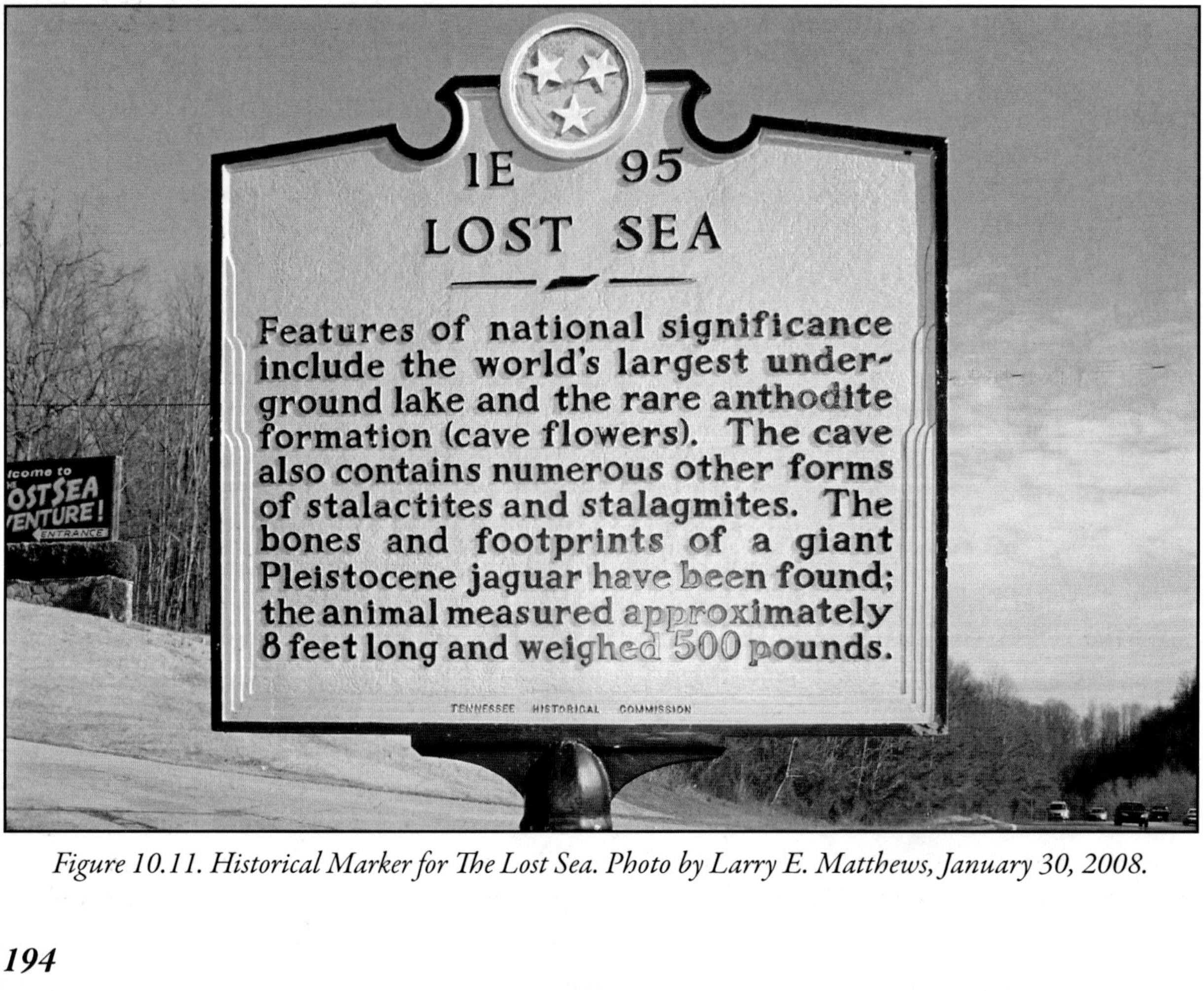

Figure 10.11. Historical Marker for The Lost Sea. Photo by Larry E. Matthews, January 30, 2008.

Figure 10.12 Lisa McClung on the trail to the Historic Entrance. Photo by Larry E. Matthews, March 12, 2008.

James X. Corgan and Emanuel Breitburg, in their book *Tennessee's Prehistoric Vertebrates* (1996), point out a significant difference between the jaguar fossils that have been found here in Tennessee and the modern jaguar of Central and South America. Tennessee's Pleistocene jaguar, *Panthera onca augusta*, weighed an estimated 450 to 500 pounds, whereas modern jaguars, *Panthera augusta*, reach only 225 pounds.[5] A larger jaguar could hunt larger prey. This difference in size is consistent with recent geologic history, since most modern mammals' ancestors were significantly larger during the Ice Ages.

One thing that we know about the habits of modern jaguars is that their favorite prey is a group of pig-like Central and South American animals known as peccaries. Fossil peccaries are also common in Tennessee, so perhaps they were the jaguar's favorite prey in Tennessee. Tennessee jaguars would have had a wide array of animals to feed on, including the juvenile elk that was found with the Lost Sea jaguar.

It is generally believed that jaguars survived in Tennessee until near the end of the last ice age, approximately 12,000 years ago. No jaguar bones have shown up in archaeological sites, so it seems probable that they were gone before American Indians arrived here, which also was about 12,000 years ago. Perhaps one of the earliest American Indians saw one of the last jaguars. We will never know.

In historic times, jaguars have been seen as far north as southern Arizona, southern New Mexico, and western Texas, where small peccaries known as javelina live. Like most big predators, jaguars were shot and poisoned by both ranchers and the federal

5 James X. Corgan and Emanuel Breitburg, *Tennessee's Prehistoric Vertebrates*, Bulletin 84, Tennessee Division of Geology, 170 pages.

Figure 10.13. Lisa McClung at the Historic Entrance, which is still used for the Wild Tours. Photo by Larry E. Matthews, March 12, 2008.

government. In the past few years, a few isolated individual jaguars have been spotted, and even photographed, crossing the border from Mexico into the United States. Perhaps, with protection under the Endangered Species Act, the jaguar will return to its historic range in the American Southwest.

A Description of the Cave

The Lost Sea is one of the premier tourist attractions in east Tennessee and tours are given daily. The first comprehensive description of the cave was published in the book *Caves of Tennessee*, which was written by Thomas C. Barr Jr in 1961. He gives the following written description of the cave:

> Craighead Caverns is one of the largest and most extensive caves in east Tennessee and one of the most unusual. It has been known for more than 100 years and is said to have been named for an Indian. The present entrance is an artificial one, which opens into the cave about 50 feet lower than the natural entrance.
>
> The main channel of the cave trends southwest along the strike for half a mile. At the northeast end it terminates in breakdown at the edge of the ridge. At the southwest end it becomes choked by clay fill. In the vicinity of the mouth is an extensive and complex lower level consisting of a network of fissures and pits. It may be entered at several points in the main cave. The main cave averages 100 feet in width and 30 feet in height.
>
> Near the southwest end is a huge underground lake 600 feet long, 200 feet wide, and as much as 65 feet deep. It is fed by two streams and is 250 feet below the elevation of

Figure 10.14. Old photograph of persons in the cave. Photo courtesy of The Lost Sea.

Figure 10.15. Old photograph of a large group on the wild tour in the cave. Photo courtesy of The Lost Sea.

Figure 10.16. The Visitor Center for The Lost Sea. Photo by Larry E. Matthews, January 30, 2008.

the [historic] entrance. It has been known to fluctuate as much as 13 feet, and in the late winter and spring it is inaccessible.

Also near the southwest end, in a section known as the "New Room," is a remarkable display of anthodites, clusters of tubular calcite needles radiating outward from a common center. Hundreds of anthodite clusters may be seen in this area.

Simpson (1941) described bones and tracks of a large Pleistocene jaguar from Craighead Caverns. Well-preserved tracks in soft mud were discovered near the New Room, and fragments of a jaguar skeleton were unearthed in part of the lower levels.

Craighead Caverns is currently open to the public from time to time and is electrically lighted.[6]

6 Thomas C. Barr Jr, *Caves of Tennessee*, Bulletin 64, Tennessee Division of Geology, 1961, pp 322–324.

A map of Craighead Caverns that was prepared by Berlen C. Moneymaker in 1929, with additions by W.E. Michael in 1940, is shown as Figure 10.2. Tom Barr recalls one of his visits to the cave in the 1950s:

> Back in the 1950s when we took one of our infamous rubber rafts into The Lost Sea during its transformation from Craighead Caverns into The Lost Sea, the water was crystal clear, though it was so deep that it was impossible to see the bottom. With only one rubber raft, I was elected to launch out onto the lake and try to see what was on the far side. We set our two gasoline lanterns on the shore, and I took off in the raft, my only light a carbide lamp (I did have the requisite spare flashlight in a pocket). There are no other landing places anywhere around the lake. When I reached the far side, about 600 feet from shore, I noted the two gas lanterns

back where I started from and called back to find out how far apart they were. I had guessed they were about 3 feet part, but the answer came back: 10 feet! Two pools of light in a big, black void.

Another tale about The Lost Sea (or Craighead Caverns, as it was once known during a brief, earlier period of commercialization) was told to me by Dr George Gaylord Simpson, mammalian paleontologist and evolutionist extraordinaire. He was my advisor when I was a graduate student at Columbia University; he was an adjunct professor of zoology at Columbia even though his office was at the American Museum of Natural History some 40 blocks downtown by subway. A short, plump, bald man with a fringe of red hair and a red goatee, he could fall completely silent if he had nothing to say to you, but once a topic that interested him was brought up, he was a great raconteur. When footprints of an apparent Pleistocene jaguar were found in Craighead Caverns around 1940, he traveled to Sweetwater to look for fossil bones. His guide took him through an extensive series of high, narrow canyons developed as side passages to the main borehole; since these all narrowed at the top and bottom, the only accessible part of the canyon was in the middle, and locomotion was by means of hands and feet pressed against one wall and buttocks pressed against the other. When they exited the cave, Simpson learned that his guide was an epileptic. He described the tracks and a few bones in 1941 but didn't find enough material to fully diagnose a new species of big cat.

Although additional jaguar remains haven't been found in The Lost Sea, they would at least be a natural element of the fauna that could have been exploited in commercial interpretation of the cave. A few years later, Ned McCrady, Harvey Templeton, and Henry Kirby-Smith discovered bones of a Pleistocene jaguar in Salt River Cave, and McCrady described

Figure 10.17. Photograph showing a person in a small rubber raft on the Lost Sea, about 1960. Photo by Thomas C. Barr Jr.

Figure 10.18. Front and back of an early brochure for The Lost Sea, showing boats on the Lost Sea. From the Speleo Archives of Gary K. Soule.

the find as a new extinct species, *Panthera augusta.*[7]

So far as I have been able to determine, Tom Barr was the first person to ever take a boat down to the Lost Sea and explore the lake. He made the following comment about being alone in the tiny rubber raft in the middle of that huge lake: "I felt quite alone in my rubber raft out in the middle of that thing."[8]

Currently, The Lost Sea has a surveyed

7 Thomas C. Barr Jr, Tag-Net Digest #4679, August 3, 2007.

8 Thomas C. Barr Jr, Personal communication, August 8, 2007.

Sail into a lost world of enchantment —

4½ acres of bottomless mystery and magic

Your ride over the placid waters of the Lost Sea is powered by electric motors so smooth that you get the sensation of silently sailing into the past! The Lost Sea contains over 30,000,000 gallons of crystal clear water ranging in depth from 10 feet to over 100 feet. Early this year Scuba divers exploring the far end went under the cliff wall and discovered a much larger subterranean lake feeding the Lost Sea. This lower level lake is estimated to contain over 1 billion gallons of water, and the bottom is yet to be found.

Frequently, huge rainbow trout—the largest in Eastern America—come out of the depths and may be seen through the glass bottom boats.

Come as you are to Craighead Caverns. The temperature remains a pleasant 58° the year around. And there are no stair steps to climb! Wide, clean, dry walkways provide a gradual slope to the many interesting rooms, some of which are among the widest, highest and largest rooms of any cavern in the Southeast. From the Prehistoric Room to the Lost Sea every second of the 55-minute trip is a fascinating experience you'll never forget, so be sure to bring your camera. There is no delay—conducted tours every few minutes (special educational rates are granted for groups.) Open daily throughout the year beginning June 7, 1965. For more information write or call—Manager, Lost Sea, Sweetwater, Tenn. Phone 337-7126.

Emerald Falls drops out of a picturesque crevice 700 feet inside the cavern, then cascades into the main gorge to form one of the streams feeding the Lost Sea. Although the water is crystal clear a greenish-grotto mist mysteriously persists in hovering over the limpid pool.

See where Giant Jaguars roamed

Bones and tracks of a large Pleistocene jaguar were discovered in one of the immense rooms of Craighead Caverns. The skeleton was unearthed and carefully removed to the American Museum of Natural History. However, the well-preserved tracks were permanently encased in plaster of Paris, and an authentic replica of the cat remains at the mouth of the crevice into which he had fallen in the Prehistoric Room.

More to see and do . . .

Huge Rainbows roam the waters of the Lost Sea.

Saltpeter Mine secretly used by Civil War troops for gunpowder.

Figure 10.19. Inside pages of an early brochure for The Lost Sea, showing a waterfall along the tourist trail. From the Speleo Archives of Gary K. Soule.

length of 7,530 feet, or 1.43 miles (2.29 kilometers). The commercial entrance to the cave is an artificial entrance that was drilled into the hillside in 1964. The cave opened to the public as The Lost Sea in June,1965.

Cave Divers Explore The Lost Sea

Expert cave diver Jim Wyatt was a member of one of the groups of persons who explored the Lost Sea in the 1970s. Below is the record of that exploration, as written by Jim Wyatt:

> The Lost Sea is listed in the *Guinness Book of World Records* as the largest underground lake in the world. Located in Sweetwater, Tennessee, about half way between Chattanooga and Knoxville, the Lost Sea was always calling those of us cave

Figure 10.20. A page inside a recent brochure for The Lost Sea. From the Larry E. Matthews collection.

divers who lived nearby to go diving.

In 1974–1975 several cave divers were negotiating with the Chairman of the Board of Directors of The Lost Sea to allow us to begin exploration of the underwater portion of the cave. Primarily Tom Mount was the "name" that gained us access to the Lost Sea. He, along with Bill Schenck, a former National Association for Cave Diving (NACD) instructor convinced the Board that we had the experience and ability to explore this system. The Board was excited for us to do this because they too were curious where the cave leads to.

The Lost Sea is a world-class tourist attraction with tours of the dry cave ending with a boat ride on the lake. The underwater caves are simply an extension of the dry cave areas. The dry cave is extensive. We spent many weeks crawling around the dry cave and caverns getting a feel for how the system was laid out before we made our first dives. During the winter months while we were exploring we stayed overnight in an equipment room near the lake and had full access to the dry cave.

Figure 10.21. An early "Sail the Lost Sea" Patch. From the Speleo Archives of Gary K. Soule.

The walk into the cave, to the lake, is about a quarter of a mile down an incline. There are no steps but carrying double 100s [scuba tanks] up and down that incline for filling was a challenge. We had an old compressor on a trailer owned by Bill Schenck, of course it was

Figure 10.22. An early small bumper sticker for the Lost Sea. From the Speleo Archives of Gary K. Soule.

Figure 10.23. Front of a recent brochure for the Lost Sea, showing tourists in an electric-powered boat. From the Larry E. Matthews collection.

powered by a gasoline engine and pumped about 8 cubic feet per minute. Some days Lost Sea staff hauled our tanks back and forth on an electric golf cart. Other days we walked them up and down the hill.

During our initial preparations for exploring the cave we had surveyors to get into the dry cave and then onto boats and made a survey of the lake. The lake was measured to be 800 feet long by 220 feet wide, about 4.5 acres. Even before the surveys we knew there were potentially several passages that would lead off into "going cave" and our initial task was to determine where first to dive.

Part of the tour that tourists get is a glass bottom boat ride onto the lake. The lake was stocked with large rainbow trout and the perimeter of the lake was lined with underwater lighting. We used one of those boats to cruise around the lake and we did our initial surveys with wet suits and mask, fins, and yes, snorkels. The water temperature was, as I recall, very cold, 52 degrees.

As chairman of the diving committee I was the record keeper, map keeper, and point of contact with the Board of Directors. All my records were in my parent's home when it caught fire, burning all my dive gear and all records and maps of our exploration, some years after the project ended. I lived 45 minutes from The Lost Sea and had recently been certified as NACD instructor #28.

There were several of us who were diving over the almost two years that we worked in that system. There would be weekends that we dived all day and night and then weeks when none of us could make it up there to dive. We required all divers to be NACD certified divers. The National Speleological Society Cave Diving Section was yet to be formed. NACD at that time had only one level of certification, Cave Diver.

Some of the divers that assisted were Jim Fishback, Victor Sparks, Sheck Exley, Mac Fair, Jeff David, Billy Young, and Tom Mount. Tom got us in the door and made a few dives with us doing the initial surveys into the underwater cave system and really got us on track exploring the main passage. Tom was at that time the training director for the NACD and his help was key. Tom lived in Miami at the time and found the 20-hour drive prohibitive to be with us every weekend that we dived. Sheck talks briefly about the Lost Sea in *Caverns Measureless to Man*, but was not able to make that long drive from north Florida very often.

Figure 10.24. Lost Sea Wild Cave Spelunker Patch. From the Speleo Archives of Gary K. Soule.

We had no way of knowing what depths we would encounter, but we all suspected that the depths would be in excess of 100 feet very shortly after beginning the exploration. We were correct. The underwater section of the cave angled down just as sharply as the dry cave did. We were diving on air, and decompressing on air. That was all we had in those days. The hi-tech gear we had included a Benjamin valve (dual outlet), harnesses attached to our tanks, and a horse collar buoyancy control device with optional power inflator. Victor Sparks elected to use oral inflation, rather than the power inflation option. Our lights were Frank Martz type lights with wet NiCad batteries and sealed beam halogen 100-watt bulbs. We got an hour burn time out of these lights. Our reels were of a Frank Martz design also. All of my reels and lights were built by Bill Schenck who never dived with us after all the work of getting access due to personal reasons not to do with cave diving.

Our primary tie off was in open water. We tied to one of the underwater lights dangling down about 10 feet under the surface. The secondary tie off was quite another problem, as was all other tie offs. We discovered immediately that all outcroppings from the wall and ceiling were very delicate and any attempt to tie line to these outcroppings resulted in destruction of the outcropping and a corresponding decrease in visibility.

The bottom was very thick with very fine red clay silt. It was in some places 6 feet thick and others as much as 2 to 3 feet thick. Nowhere that we ever found had a rocky bottom. We resolved the no-tie-off-point issue with stakes we would drive into the silt. This silt was almost like Jell-O, in that it wiggled when you got near it.

These stakes we made out of reinforcing bar with a fender washer welded onto the top. Mac Fair volunteered to make up a batch of these stakes before our next series of dives. On our first set of dives we installed no line because of this.

Our next trip up, armed with our exploratory reels and homemade tie off

Figure 10.25. Cave diver Jim Wyatt. Photo courtesy of Jim Wyatt.

Figure 10.26. Anthodites in The Lost Sea. Photo by Larry E. Matthews, March 12, 2008.

points, we attempted to install exploratory line into the system. Mac Fair made these stakes for us before every trip up to dive. We decided on three-man teams for line installation because the weight of the stakes was so much that is was too much for one diver to carry while one diver was laying line. One diver ran a reel, and both other divers would carry one or two of these stakes.

We installed a lot of line that next weekend. The stakes worked fine, they were all 6 feet long and most of them went into the silt to a point where only a foot to 18 inches stuck out. The diver placing the stake had to be extremely careful and be slow in placing the stake so as not to completely blow out the 15 feet of visibility we had most of the time on the way in.

Jim Fishback was employed by the U.S. Geological Survey during the time we were doing this exploration. One weekend he brought with him an underwater sonar unit so we could measure dimensions of the rooms we were swimming through. Most of the time we were unable to get any sense of the size of the underwater rooms due to the limited visibility as well as the enormous size of the rooms and the cave.

This unit was a flashing type fish finder in a waterproof housing. It would read as deep as 120 feet if it were mounted on a boat. We took this unit with us on dives after we had installed line and were attempting to make an initial map of the cave. The unit maxed out at 120 in the first room we came to after the lake. We were along one wall, but the room was so large that we could not get a reading on the opposite wall or the end of the room. The room was at a point of about 400 feet of penetration and 110 to 120 feet of depth. It had taken us several dives to install 400 feet of line.

We decided to continue installing line as we had leveled off at 110 to 120 feet

Figure 10.27. Modern billboard for The Lost Sea. Photo by Larry E. Matthews, March 12, 2008.

and the cave was going. One dive I recall coming to the end of my reel, after having installed about 300 feet of line. We drove a stake into the silt and I tied off the end of the line to the fender washer. We blew the visibility out to zero, and as I was cutting the line and stowing my empty reel I realized I was off of the line and I had no safety reel. I knew I was in deep trouble if I did not relocate that line soon. There was no way to wait out the clay silt, there was no visibility, and I would not expect my buddy to be able to find me. He had no safety reel either. Safety reels were introduced to cave divers after this event took place and we were not trained to use them, or even knew of their existence. Obviously I found the line after a minute or two (seemed like 20). Using the methodical, non-line search I had been taught in my 6-month-long, full-cave course.

We found anthodites or cave flowers at 65 feet deep on the walls. This was fascinating to the geologists and other science guys because we know that these cave flowers cannot form underwater. These flowers were photographed, copies sent to the Board, and others burned in the house fire that also took my high school annuals and other memorabilia of my 20 years on Earth.

We installed one continuous guideline, with no off-shooting lines, because the passage kept going. Overall, we installed about 1,500 feet of line in that system.

The one event that continually occurred made us all think long and hard about continuing this exploration was the percolation that we experienced on every dive. Since no human had ever been in there before us, the overhead had clinging to it all the silt of a thousand years or more, and every time we exhaled stuff rained down on top of us. Mostly it was clouds

Figure 10.28. Entrance Sign for The Lost Sea property. Photo by Larry E. Matthews, March 12, 2008.

of silt, but occasionally a few small rocks would fall.

During every exit we experienced 5 feet or so of visibility due to this percolation. Couple that with 100 to 120 feet of depth on air, 52 degree water, and it created an added stress factor for us all.

One dive with Tom Mount as team leader, and Victor Sparks as team member, an event took place that made us re-evaluate the safety of the project. Tom and Victor described a large boulder falling from the ceiling and almost hitting them. As I recall, they stated that the rock, the size of a Volkswagen beetle, could have pinned them to the bottom with little hope of escape. The Chairman of the Board asked for my word, early on in our exploration, that I would shut the project down should we find conditions that we felt were more risky than we should accept as a risk factor.

Many of us decided, after this incident, that things on the ceiling were loosening up because of our bubbles. More and more rocks and debris were falling down on us on every dive. We knew we had explored about 1,500 feet of passage, created a map and discovered significant water table changes. We saw changes in the geology with the depth and were able to relate those changes to various geologists who learned from it. We did not know where the cave ended up. We never walled it out.

Because we were diving and decompressing on air we were spending long hours in that frigid water. The dives were getting longer and longer with each push and the decompression was getting colder and colder.

Most of us felt we had reached our limits and I notified the Chairman of the Board, provided them with maps and a

Figure 10.29. "The Lost Sea Adventure" sign. Photo by Larry E. Matthews. January 30, 2008.

few photographs, and ended the project with no fatalities. There was a couple in our group who felt we should continue pushing the cave and we will never know if we called it off too soon. I sleep well at night making the decision to stop.

As far as the future for further exploration of the Lost Sea there is great potential for closed circuit re-breather explorers, with dry suits and modern decompression gasses. I think the property was sold to a group of attorneys who have expressed reluctance for further exploration due to liability issues. I have been contacted a couple of times by divers seeking advice on how to get back in there. So far no one has seriously followed up on it.

We explored only one passage off of the lake. It kept going and going and we saw no need to look for other passages. My guess is that there are many more passages waiting to be explored.

As Jim Wyatt notes, they explored the Lost Sea in the early days of cave diving technology. In all likelihood, the cave could be explored much further with modern cave diving equipment. However, commercial caves cannot operate without insurance, and it would be difficult, at best, to get permission to explore an underwater cave passage in today's legal climate.

The Lost Sea Today

Today, The Lost Sea is one of the major tourist attractions in the Knoxville/Great Smoky Mountains National Park area. Each tour lasts approximately one hour. The highlight of the tour is a glass-bottomed boat ride on the huge underground lake, which is the largest in the United States. There are no steps in the cave, but there is a gradual incline as you walk back to the entrance from the lake. Since the cave is a constant 58 degrees Fahrenheit, you will want to wear a light jacket or a sweater. Comfortable walking shoes are also advisable.

Also available are **Wild Cave Tours**, but these must be scheduled in advance. They are ideal for organized groups, such as Scouts, school groups, and church groups. Persons going on the Wild Cave Tour will take the regular cave tour as well as an exciting tour of the undeveloped portions of the cave where they will be required to climb and crawl through the natural cave passages. The wild tour includes an overnight stay in the cave. Each person will need to bring a flashlight, extra batteries, a canteen for drinking water, a sleeping bag, and an air mattress or ground cloth. There are no showers or running water in the cave, but there is a toilet. You will get muddy, so be sure to bring a complete change of clothes to change into when you leave the cave.

Figure 10.30. Front of a brochure for the Cherohala Skyway. From the Larry E. Matthews collection.

The cave is open every day of the year for tours, except Christmas Day. For more information you may telephone:

423-337-6616

Or, you can send e-mail to the following address:

lostsea@earthlink.net

The mailing address for The Lost Sea is:

The Lost Sea
140 Lost Sea Rd
Sweetwater TN 37874-6724

Directions

The Lost Sea is an easy, one-hour drive from Knoxville. From Knoxville, go west on I-40 for 18.7 miles to the junction with I-75. Exit onto I-75 and proceed southwest for 24.0 miles to Exit 60 (Sweetwater/Spring City Exit). Turn left onto Highway 68 and proceed southeast for 7.5 miles. Turn left onto Lost Sea Road and proceed 0.1 mile to the Visitor Center. There is a large sign at the intersection of Highway 68 and Lost Sea Road.

Side Trips

If you have the time, continue southeast on Highway 68 an additional 19 miles to Tellico Plains and the entrance to the Cherohala Skyway. The Cherohala Skyway extends east for 36 miles to Robbinsville, North Carolina. For the first 21 miles you will be climbing through the Cherokee National Forest to the crest of the Great Smoky Mountains. At this point, you will cross into North Carolina and for the

Figure 10.31. A young person exiting from a crawlway on the Wild Tour in The Lost Sea. Photo courtesy of The Lost Sea.

next 15 miles you will be descending through the Nantahala National Forest. For the entire 36 miles you will be rewarded with spectacular views of the mountains while driving on a modern highway. There are numerous overlooks and well-marked hiking trails along the way. This drive is as beautiful as any drive in the Great Smoky Mountains National Park, but with far less traffic.

Chapter 11

Tuckaleechee Caverns

The Discovery of Tuckaleechee Caverns

The area where Tuckaleechee Caverns is located was in the heart of Cherokee Indian Territory. However, there is no evidence that the Cherokee, or any previous prehistoric people living in this area, ever entered and explored the cave. Tuckaleechee is a Cherokee word that translates as "green, peaceful valley." Tuckaleechee Cove is still a beautiful place and well worth a visit.

Lumbermen discovered what is now known as Tuckaleechee Caverns when this area was being logged for its virgin timber. These original explorers used pine torches and kerosene lanterns to explore the stream passage. This was in the late 1800s, after the railroad had arrived and the wood could be shipped off to distant markets.

The caverns were first opened to the public for a year in 1931, but were soon closed because of the Great Depression. No one had enough money to travel or buy tickets. The cave was known as Great Smoky Mountain Cavern at this time.

As young boys, W.E. "Bill" Vananda and Harry Myers, both played near the entrance to the caverns and frequently ventured into it. When they were college students at nearby Maryville College in 1949, they began to discuss the possibility of opening the cave to the public. Soon they were building trails through the cave. After four years of labor, the cave opened to the public in 1953. These early tours used Coleman lanterns for light. In 1955 electric lights were installed in the cave.

In 1954 Bert H. Denton visited the cave as a part of the Tennessee Division of Geology's Cave Survey. While in the cave, his group discovered what is now known as the Big Room. This room is more than 400 feet long, 300 feet across, and 150 feet deep, with stalagmites up to 24 feet high. Vananda and Myers soon built trails into this new discovery and it was opened to the public in 1955. At the same time, they installed electric lights, eliminating the use of kerosene lanterns, which had been used for the first two years.

On April 9 through 12, 1958, the National Speleological Society held its 15th National Convention in nearby Gatlinburg and held a banquet inside Tuckaleechee Caverns. The NSS was a small group in these early days, and only a little over 70 members were present for the underground banquet. John and Norma Wilson, of Wilson's Hillbilly Restaurant, served the meal (southern style country ham and fried chicken with all the fixin's) on tables with white tablecloths inside the cave. During the week of the Convention, a cave diver explored the sump in Tuckaleechee Caverns. A photo by Leonard J. Munson in the May 1958 *NSS News* shows the diver entering the sump, but does not give the diver's name.[1]

Bill Vananda and his wife, Golden, along with Harry Myers and his wife, Nita, owned and operated Tuckaleechee Caverns from 1953 until 1982. In an interesting arrangement, each couple ran the business on alternate days from April through October, seven days a week. In 1982 the Myers sold their interest in the Caverns to the Vanandas, who still own the cave.

---

1 Anonymous, "National Speleological Society 15th Annual Convention – Gatlinburg, Tennessee," *NSS News*, vol 16 (1958), no. 5, pp 41–47.

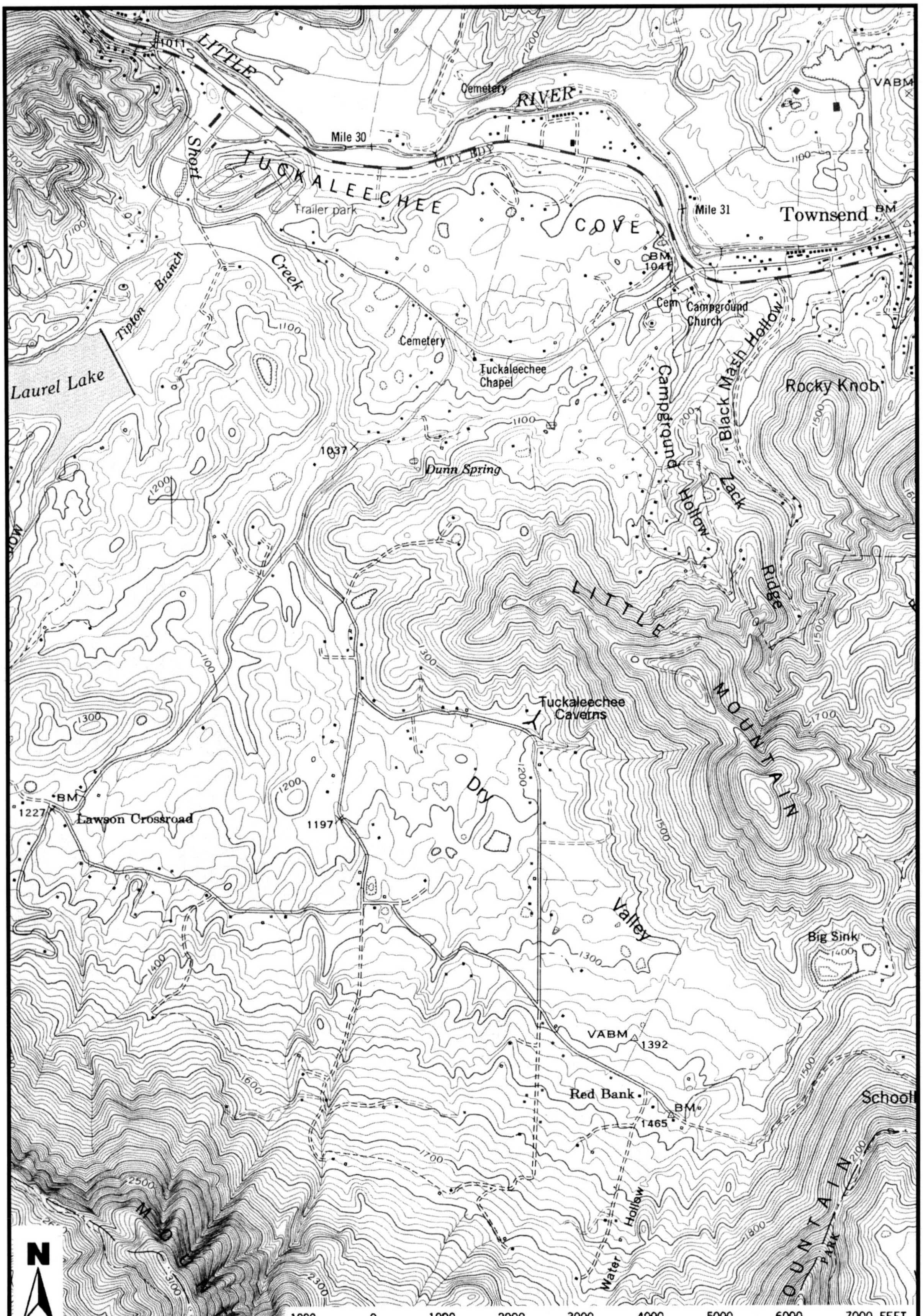

Figure 11.1. Topographic map showing the location of the entrance to Tuckaleechee Caverns.

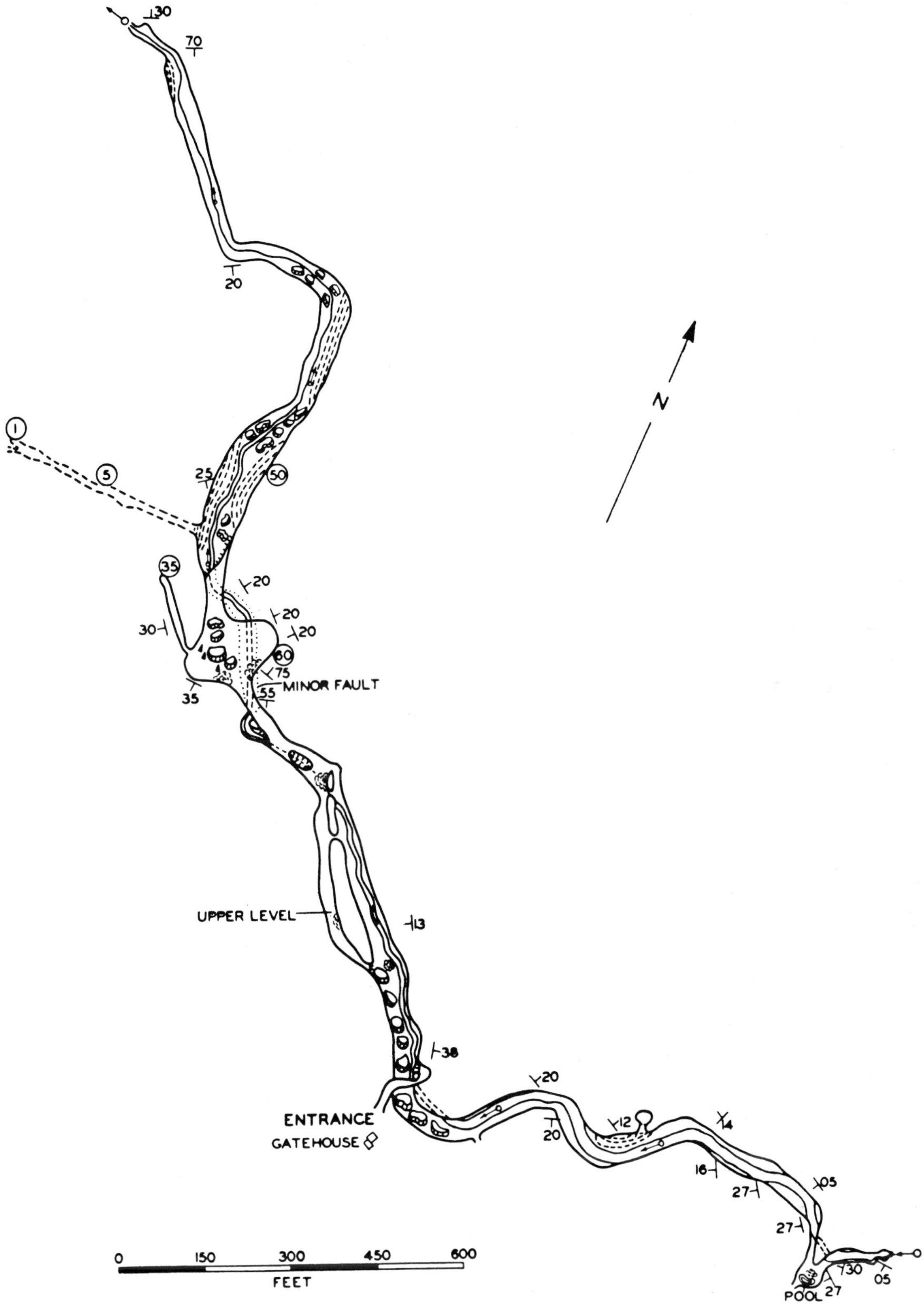

Figure 11.2. Map of Tuckaleechee Caverns from Caves of Tennessee, *page 83. T.C. Barr and B.C. Stewart, 1956.*

NATIONAL
SPELEOLOGICAL
SOCIETY
BANQUET

April 9, 1958 - 6:00 P.M.

TUCKALEECHEE CAVERNS

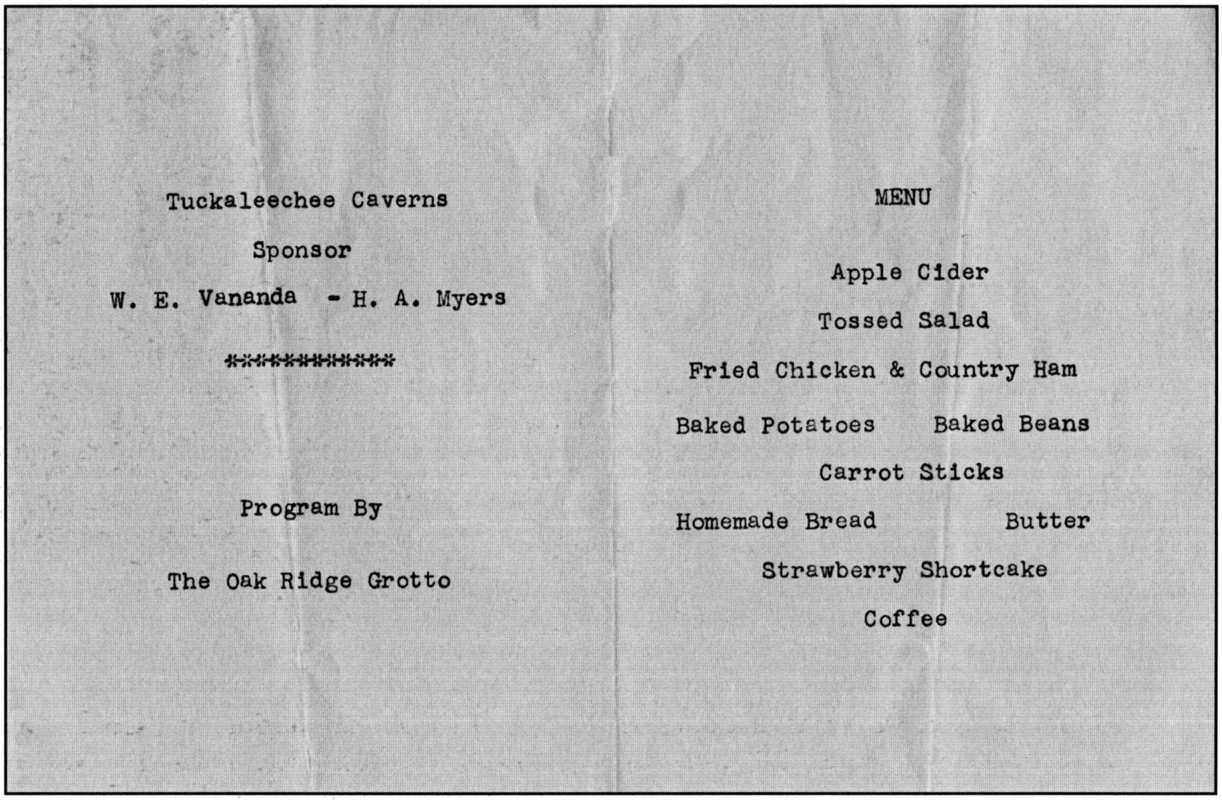

Tuckaleechee Caverns

Sponsor

W. E. Vananda - H. A. Myers

****

Program By

The Oak Ridge Grotto

MENU

Apple Cider

Tossed Salad

Fried Chicken & Country Ham

Baked Potatoes Baked Beans

Carrot Sticks

Homemade Bread Butter

Strawberry Shortcake

Coffee

Figure 11.3. The Banquet Menu from the 15th Annual National Speleological Society Convention, April 9, 1958. From the Gordon and Judy Smith collection at the National Cave Museum, Diamond Caverns, Park City, Kentucky.

Figure 11.4. Front and back of the original Tuckaleechee Caverns brochure. "Nature's Underground Wonderland." From the Speleo Archives of Gary K. Soule.

When Tuckaleechee Caverns first opened in 1953, they had approximately 2,000 visitors per year. The cave's popularity has continued to grow and they now host over 50,000 visitors per year.

Tuckaleechee Caverns was designated as a fallout shelter during the Cold War, when Civil Defense officials determined that Tuckaleechee Caverns could handle 3,500 people in case of a nuclear attack. Fortunately, Tuckaleechee Caverns never had to be used as a fallout shelter and caves are no longer

Figure 11.5. Inside of the original Tuckaleechee Caverns brochure. "Caverns so Large They are Still Being Explored!" From the Speleo Archives of Gary K. Soule.

considered a suitable site for such use.

A Description of the Cave

Tuckaleechee Caverns is one of the main tourist attractions in the vicinity of the Great Smoky Mountains National Park. The book, *Caves of Tennessee*, which was written by Thomas C. Barr Jr in 1961, gives the following written description of the cave:

> Tuckaleechee Caverns is the largest known cave in Blount County. Part of the

Figure 11.6. Early photograph of a drapery formation in Tuckaleechee Caverns from Caves of Tennessee, *page 84.*

from the entrance for 1,100 feet and northwestward for 2,100 feet. Additional smaller passages total 1,400 feet. The total explored length of the caverns is thus about 4,600 feet.

An upper level, which may be traced along part of the downstream section, culminates in a huge room 50 feet above the stream. This room—which is 165 feet long, 100 feet wide, and 60 feet high—probably had never been entered before September 9, 1951, when it was discovered and explored by Harry Myers (one of the caverns operators), Bert Denton, and the writer [Barr]. Several attractive displays of dripstone are found in this chamber, including two large stalagmites 18 inches to 2 feet in diameter and 18 feet tall. A tight hole at the west corner of the room opens into a 50-yard passage that contains some of the most beautiful formations in the cave.

Downstream, beyond the big room, the main cave expands into a large gallery 70 feet wide, 50 feet high, and 350 feet long. Beyond this point the cave averages 25 feet wide and 15 to 20 feet high.

Breakdown is extensive in the vicinity of the entrance and downstream well beyond the large gallery. Dripstone is

cave is electrically lighted and is exhibited to tourists. The entrance is in the bottom of a small sinkhole, and the entrance passage crosses over the main cave, which it joins 40 feet above the floor. A sizable stream flows through the caverns, emerging from a siphon upstream and sinking beneath the floor at the downstream end. The stream is thought to originate in White Oak Sink, 2 miles southeast of the caverns entrance, and probably emerges at Dunn Spring, three quarters of a mile northwest of the entrance.

The main cave passage extends eastward

Figure 11.7. Early photograph of draperies and flowstone in Tuckaleechee Caverns from Caves of Tennessee, *page 85. Photo by Roy A. Davis.*

best developed in the downstream section as far as the upper-level room.

The Knox Dolomite, in which the cave is excavated, is exposed in the Townsend window, and in the vicinity of the caverns it exhibits a highly complex structure. The dip of the beds was determined at about 40 points within the cave, and nowhere was a consistent reading obtained for more than a few yards along the main passage. The average inclination from the horizontal is about 25 degrees. Minor flexures may be observed at several places in the walls of the cave. A fault with a displacement of a few inches may be easily recognized on the east (right) wall of the stream canyon that runs beneath the large upper-level room. The angle of dip ranges from a few degrees to 75 degrees, and the direction of dip is also highly variable, but 85 percent of the directional determinations fall within the northeast and southwest quadrants. The cave parallels the strike, crosses steeply dipping beds, and follows the axes of minor flexures for short distances, but the pattern of the whole cave is indicative of strong joint control.[2]

A map of Tuckaleechee Caverns that was prepared by T.C. Barr and B.C. Stewart in 1956 is shown in Figure 11.2.

2 Thomas C. Barr Jr, *Caves of Tennessee*, Bulletin 64, Tennessee Division of Geology, 1961, pp 82–87.

Tuckaleechee Caverns Today

Tuckaleechee Caverns is one of the most popular tourist attractions in the vicinity of the Great Smoky Mountains National Park. The cave is profusely decorated with a great variety of flowstone formations, including large totem-pole stalagmites. The tour covers approximately half a mile of cave passages.

The original entrance to the cave is now covered by a part of the Ticket Office. From this natural entrance, stairs lead steeply downward into a very large stream passage. The first half of the tour goes in the downstream direction for a quarter of a mile and terminates in an enormous room that is 100 feet wide, 160 feet long, and up to 65 feet high. In this room there are a very large shield formation and two spectacular totem-pole stalagmites, along with numerous other cave formations. After returning to the stairs descending from the entrance, the second half of the tour goes in the upstream direction for a quarter of a mile and ends where a waterfall splashes onto a large stalagmite. The entire tourist trail is concrete, but the trail is wet in places, due to dripping water, and good footwear is advised. Since the cave is 56 degrees year round, a light jacket or sweater is recommended.

For further information on Tuckaleechee Caverns, you may telephone them at:

Figure 11.8. Early photograph of draperies and a stalagmite in Tuckaleechee Caverns from Caves of Tennessee, *page 86.*

865-448-2274

The mailing address for Tuckaleechee Caverns is:

Tuckaleechee Caverns
825 Caverns Road
Townsend TN 37882-4505

You may e-mail Tuckaleechee Caverns for information at:

info@tuckaleecheecaverns.com

TUCKALEECHEE CAVERNS–A LIVING MASTERPIECE!

Long, long ago, deep under the foothills of the Great Smokies, Nature, with chisel and a palette of lavish color, created a living masterpiece. Today, a seemingly endless gallery of low ceiling passages and towering chambers are filled with intricately sculptured exotic formations, splashed in rainbow hues, glittering with star dust, and polished by the hands of Time. And within this wonderland fantasy a mountain stream has long since discovered a way to penetrate the chasm, thus contributing an added underground display of dazzling reflections.

So astounding is the spectacle that TUCKALEECHEE CAVERNS at Townsend, Tennessee, have already become a popular top ranking natural scenic attraction. Little known and practically unexplored prior to 1954, when development for public showing was first seriously and successfully undertaken, this labyrinth of spectacular underground beauty is now considered the most ancient, largest and colorful in eastern America.

Situated half way between Gatlinburg and Maryville, Tennessee, but two miles of easy motor access from Tennessee Highway 73, TUCKALEECHEE CAVERNS in a comparatively short period, have become another worthwhile educational point of interest for the thousands who annually enjoy the natural scenic wonders of Great Smoky Mountains National Park.

Through this maze of underground passages and halls, resplendent with glittering stalactites—truly a setting for the most fantastic fairy tale— the management of TUCKALEECHEE CAVERNS, from April 1st through October 31st, conducts daily a succession of guided hour-long educational tours. Lecturers are thoroughly competent and painstaking. Illuminated walkways are safe and kept in excellent repair. There is ample automobile parking space at the gatehouse.

Owned and Operated
by
W. E. VANANDA — H. A. MYERS

Figure 11.9. Inside pages of the original Tuckaleechee Caverns brochure. "Tuckaleechee Caverns—A Living Masterpiece!" From the Speleo Archives of Gary K. Soule.

Directions

From Knoxville, proceed south on U.S. Highway 129 for 17 miles to Maryville, Tennessee. Turn left (east) on U.S. Highway 321 and proceed for 14 miles to Townsend, Tennessee. Turn right onto Old Tuckaleechee Road. There is a large Tuckaleechee Caverns billboard at this intersection. Go 0.7 mile and turn left onto Old Cades Cove Road. Go 0.7 mile and turn left onto Dry Valley Road. Proceed 0.4 mile and turn left onto Cavern Road. Continue 0.4 mile and turn left into the parking lot for Tuckaleechee Caverns.

From the Sugarlands Visitor Center in Great Smoky Mountains National Park drive west on the Little River Road for 17.9 miles

Figure 11.10. Photograph of a Tuckaleechee Caverns billboard along the highway. Photo by Larry E. Matthews, October 21, 2007.

Figure 11.11. Photograph of the visitor center for Tuckaleechee Caverns. Photo by Larry E. Matthews, October 21, 2007.

Figure 11.12. Front and back of a color brochure for Tuckaleechee Caverns. "Greatest Sight Under the Smokies." From the Speleo Archives of Gary K. Soule..

to an intersection. Turn right, west, onto State Highway 73. This will lead you out of the National Park into Townsend. At 2.3 miles you will reach a junction with Highway 321. Continue straight west on Highway 321 for 1.9 miles. Turn left onto Old Tuckaleechee Road. There is a large Tuckaleechee Caverns billboard at this intersection. Go 0.7 mile and turn left onto Old Cades Cove Road. Go 0.7 mile and turn left onto Dry Valley Road. Proceed 0.4 mile, then turn left onto Cavern Road. Continue 0.4 mile and turn left into the parking lot for Tuckaleechee Caverns.

There are numerous signs and billboards leading to Tuckaleechee Caverns, so it is very easy to locate, no matter what direction you are

Figure 11.13. Photograph of the entrance to Tuckaleechee Caverns. Photo by Larry E. Matthews, October 21, 2007.

Figure 11.14. Photograph of Betty Matthews at Dunn Spring where the Tuckaleechee Caverns stream comes to the surface. Photo by Larry E. Matthews, October 21, 2007.

Figure 11.15. Front and back of a recent brochure for Tuckaleechee Caverns. "Greatest Sight Under the Smokies." From the Speleo Archives of Gary K. Soule.

coming from.

Side Trips

Between the main road (U.S. 321) in Townsend and Tuckaleechee Caverns, you will see a sign advertising a trout farm and restaurant. If you turn here, the road soon leads to Dunn Spring, where the stream in Tuckaleechee Caverns comes to the surface. This cold, clear water is being used at the present time to raise trout in a series of pools. A restaurant sits next to this spring and serves trout dinners. The spring is very beautiful and well worth a visit.

Townsend is located at one of the main entrances to the Great Smoky Mountains National Park and this is the closest entrance to

Cades Cove, one of the most popular spots in the park. From Townsend, follow the signs into the park. Almost immediately after entering the park, you will come to a "T" intersection. Turn right and follow this Laurel Creek Road for approximately 10 miles to Cades Cove. A one-way loop road will take you around Cades Cove. Gregorys Cave (See Chapter 9.) is located in Cades Cove. Cades Cove is heavily visited during the peak of the fall colors, so if you decide to visit at this time, the traffic will be extremely heavy and slow.

Figure 11.16. A recent advertisement for Tuckaleechee Caverns. From the Speleo Archives of Gary K. Soule.

Figure 11.17. A Tuckaleechee Caverns decal. From the Speleo Archives of Gary K. Soule.

Figure 11.18. A Tuckaleechee Caverns patch. From the Speleo Archives of Gary K. Soule.

Chapter 12

Appalachian Caverns

The Early History of Linville Cave

According to cave historian Jim Whidby, Linville Cave was named for two brothers who were hunters. One was buried in the cave after being killed by Indians.[1] This would probably place the discovery of the cave somewhere in the late 1700s. Linville Cave is located in Sullivan County, Tennessee.

Teeth of caribou (*Rangifer tarandus*) have been recovered from the cave. Today, caribou live in Canada and Alaska, so these teeth must date from one of the Ice Ages when Tennessee was much colder than it is today. Ironically, the remains of a giant armadillo (*Dasypus bellus*) have also been recovered from the cave. These animals lived in much warmer climates than modern-day Tennessee and must date from one of the warmer periods between the Ice Ages.[2]

A Description of the Cave

The book, *Caves of Tennessee*, which was written by Thomas C. Barr Jr in 1961, gives the following written description of the cave:

> This cave has three entrances, one on the upper level (300 feet in length) and two on the lower level (which is about 1,000 feet in length). Linville Branch enters and leaves the lower level by the two lower entrances. The stream is 4 feet deep where it enters the cave, and it leaves the cave through a lake of undetermined depth. This lake is located in the large room formed by junction of upper and lower levels. One short crawlway near the upstream entrance leads over rimstone pools to a small room containing dripstone formations. The cave trends southeast.[3]

Appalachian Caverns is much larger than the description given above and is currently being mapped by the Mountain Empire Grotto of the National Speleological Society.

The Bats of Appalachian Caverns

Appalachian Caverns is the home of a summer colony of grey bats. Grey bats frequently will spend the summer in one cave, then hibernate in another cave for the winter. At one time, grey bats were the most abundant mammals in the southeastern United States. Due to habitat destruction and multiple other causes, they are now on the Endangered Species List.

Appalachian Caverns is also inhabited by several other species of bats, including the eastern pipistrelle and big brown bats. The visitor will see some species of bat on almost every tour.

Appalachian Caverns Today

The commercial development of Linville Cave began in 1988. The original owners of the cave, Bob Joyner and Jerry and Brenda Stone, spent three years developing the cave before it was ready to open to the public. Part of this development included the removal of 162

1 Jim Whidby, Personal communication, May 23, 2007.

2 James X. Corgan and Emanuel Breitburg, *Tennessee's Prehistoric Vertebrates*, Tennessee Division of Geology, Bulletin 84, 1996, pp 141–142.

3 Thomas C. Barr Jr, *Caves of Tennessee*, Tennessee Division of Geology, 1961, p 436.

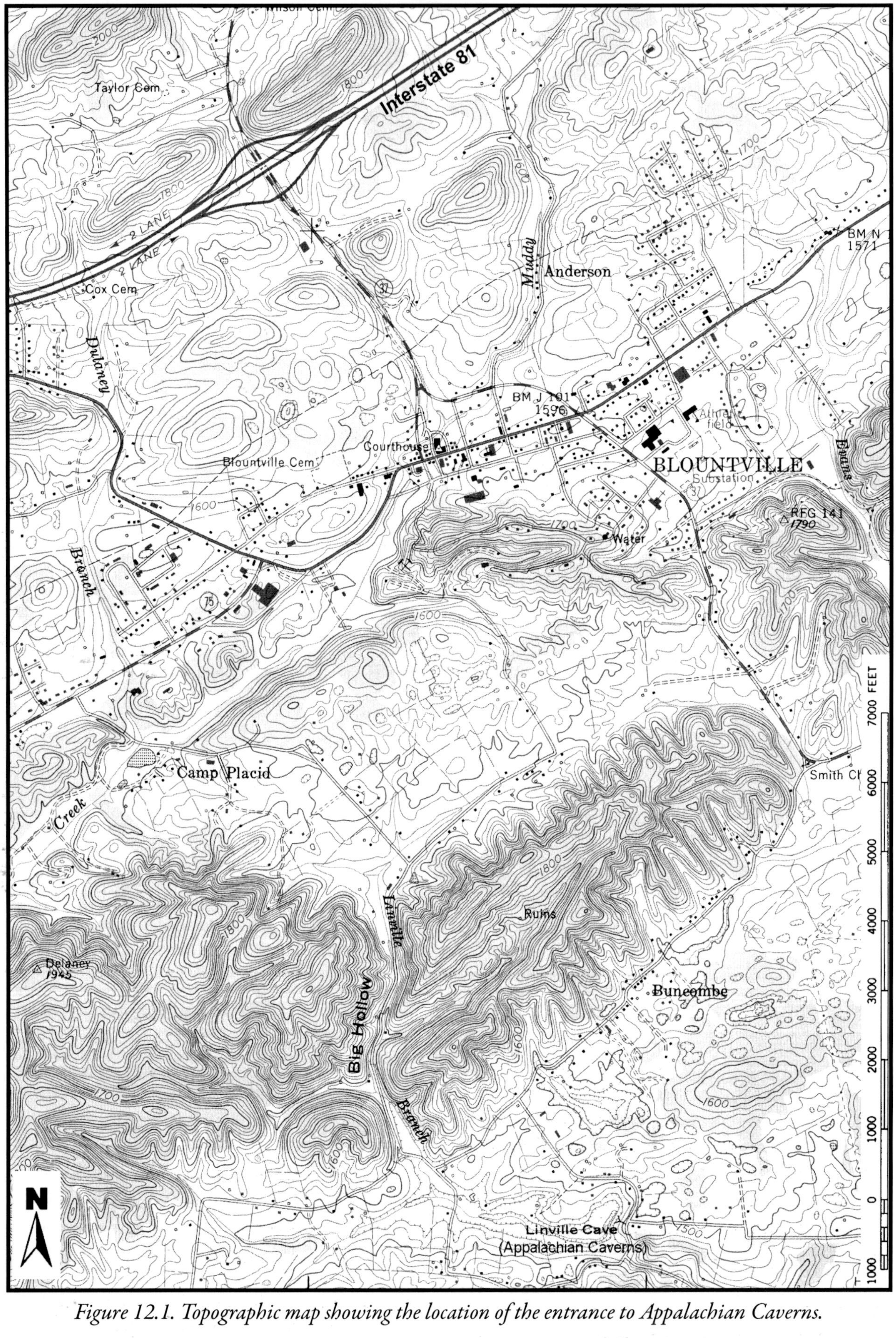

Figure 12.1. Topographic map showing the location of the entrance to Appalachian Caverns.

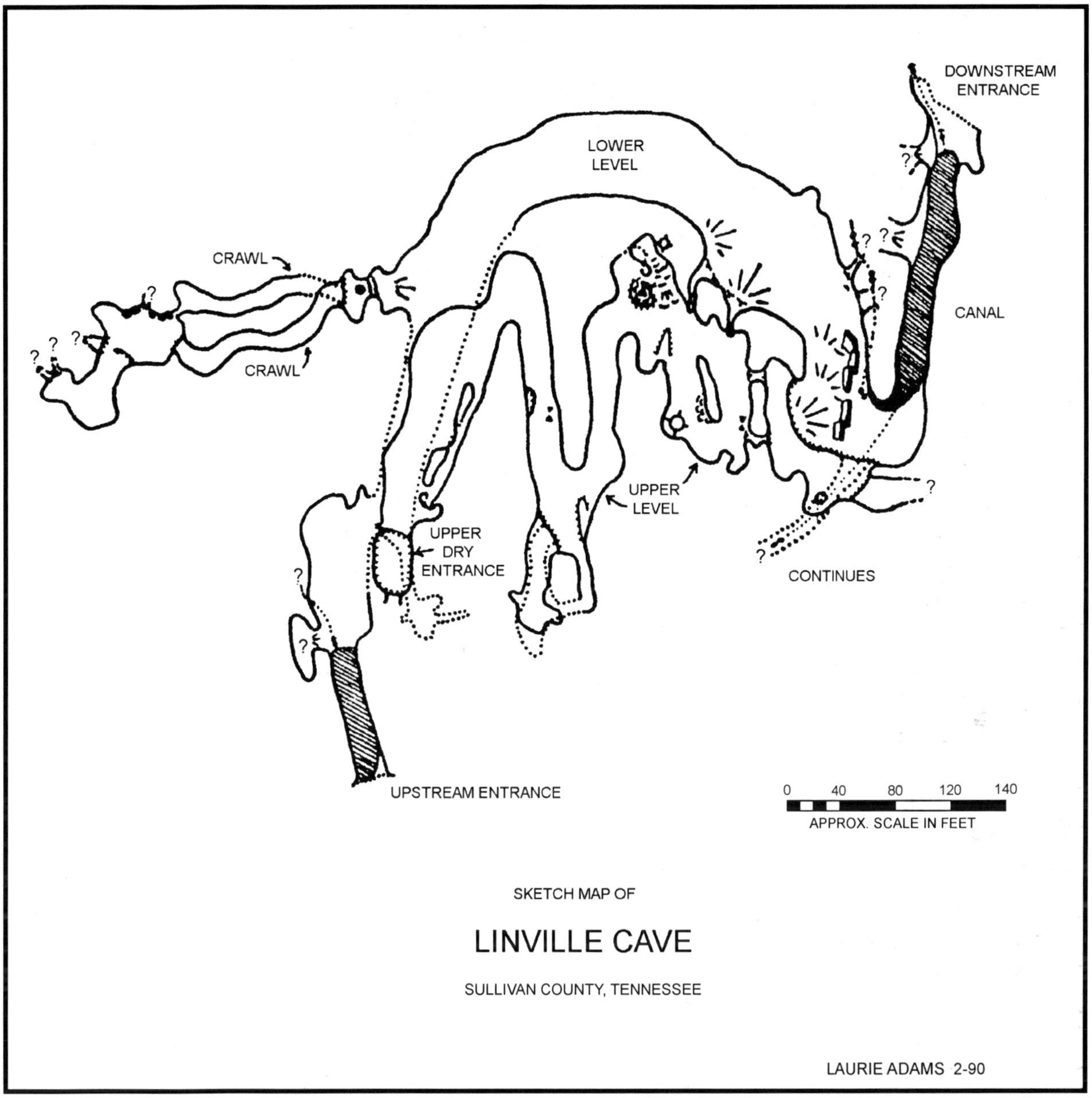

Figure 12.2. Map of Appalachian Caverns. Sketch map by Laurie Adams, February 1990.

tons of trash that had accumulated in the cave over the years. A large building was constructed over the upper entrance and this building connects to the ticket office. They changed the name to Appalachian Caverns to avoid any confusion with the commercial cave named Linville Caverns located in North Carolina. The cave opened to the public in 1991.

Appalachian Caverns closed in May 2004 and was put up for auction by the owners, Jerry and Brenda Stone. The cave was later opened under new ownership. The new owners are Roger and Susan Hartley.

The cave tour lasts approximately one hour

Figure 12.3. An Appalachian Caverns patch. From the Speleo Archives of Gary K. Soule.

Figure 12.4. The front and an inside page from the original brochure for Appalachian Caverns. From the Larry E. Matthews Collection.

and leaves the gift shop about every 30 minutes. The tour features almost one mile of walkways, which wind through a labyrinth of corridors and rooms. In some places, the roof reaches the incredible height of 135 feet. The cave features some amazing bridges, which allow the visitor to tour the cave with ease and view the passages from several different levels. The stainless steel guard rails are a work of art by themselves. The cave is 52 degrees year round, so a light jacket or sweater is recommended. Casual dress and comfortable walking shoes are also advised.

Appalachian Caverns also offers a **Wild Cave Tour** into the undeveloped portions of the cave. This trip lasts approximately 3½ hours and will introduce you to what it is like

Figure 12.5. Road sign pointing the way to Appalachian Caverns. Photo by Bob Biddix, March 23, 2008.

Figure 12.6. Betty Mattthews in front of the Ticket Office for Appalachian Caverns. Photo by Larry E. Matthews, March 20, 2008.

Figure 12.7. Appalachian Caverns sign on side of an old barn. Photo by Bob Biddix, March 23, 2008.

Figure 12.8. Erica Sughrue in the commercial entrance to Appalachian Caverns. Photo by Bob Biddix, March 23, 2008

Figure 12.9. Passage inside Appalachian Caverns. Note stainless steel guard rails. Photo by Bob Biddix, March 23, 2008.

to explore a wild cave. This trip is much more challenging than the commercial tour, but you will be led by an experienced guide and it is a safe introduction to wild caving. This tour is especially popular with school groups, scout groups, and church groups. Reservations for the Wild Cave Tour should be made in advance. Anyone going on the wild tour should bring a complete change of clothes as they will become muddy and dirty in the wild parts of the cave.

Appalachian Caverns has parking for tour buses and school buses and the picnic grounds are designed to accommodate large groups. There is also a campground with private restrooms and hot showers.

The cave is open every day of the year for tours, except Christmas day. For more information you may telephone:

423-323-2337

Or, you can send e-mail to the following address:

appacavernscom@aol.com

The mailing address for Appalachian Caverns is:

Appalachian Caverns
420 Cave Hill Rd
Blountville TN 37617-6224

Directions

From Knoxville, go east on I-40 for 34 miles to the I-81 junction. Go north on I-81 for 69 miles to Exit 69. Turn right (east) on Tennessee Highway 394 and proceed 1.6 miles. Turn right (south) onto Big Hollow Road. Go 2.4 miles and turn left onto Cave Hill Road. Proceed 0.4 miles to the Appalachian Caverns Visitor Center, which is located on the left.

The current brochure for Appalachian Caverns includes a good road map which shows

Figure 12.10. Front of the current brochure for Appalachian Caverns. From the Larry E. Matthews collection.

how to reach the cave from the west (Interstate 81) or the east (U.S. Highway 11E).

Side Trips

The world-famous Bristol Motor Speedway is located nearby on U.S. Highway 11. You may wish to schedule your visit so that you can tour Appalachian Caverns and attend a race on the same trip.

Another interesting place to visit is the Gray Fossil Museum, located at Gray, Tennessee, about 10 miles further south. Gray is located on Tennessee Highway 75, 1.8 miles north of Interstate Highway 26, Exit 13. This natural history museum opened to the public on August 31, 2007, and is open daily from 8:30 A.M. until 5:00 P.M. The street address is 1212 Suncrest Drive and the toll-free telephone number is 866-202-6223.

The state of Tennessee was widening the

Figure 12.11. Brochure for the Gray Fossil Museum.

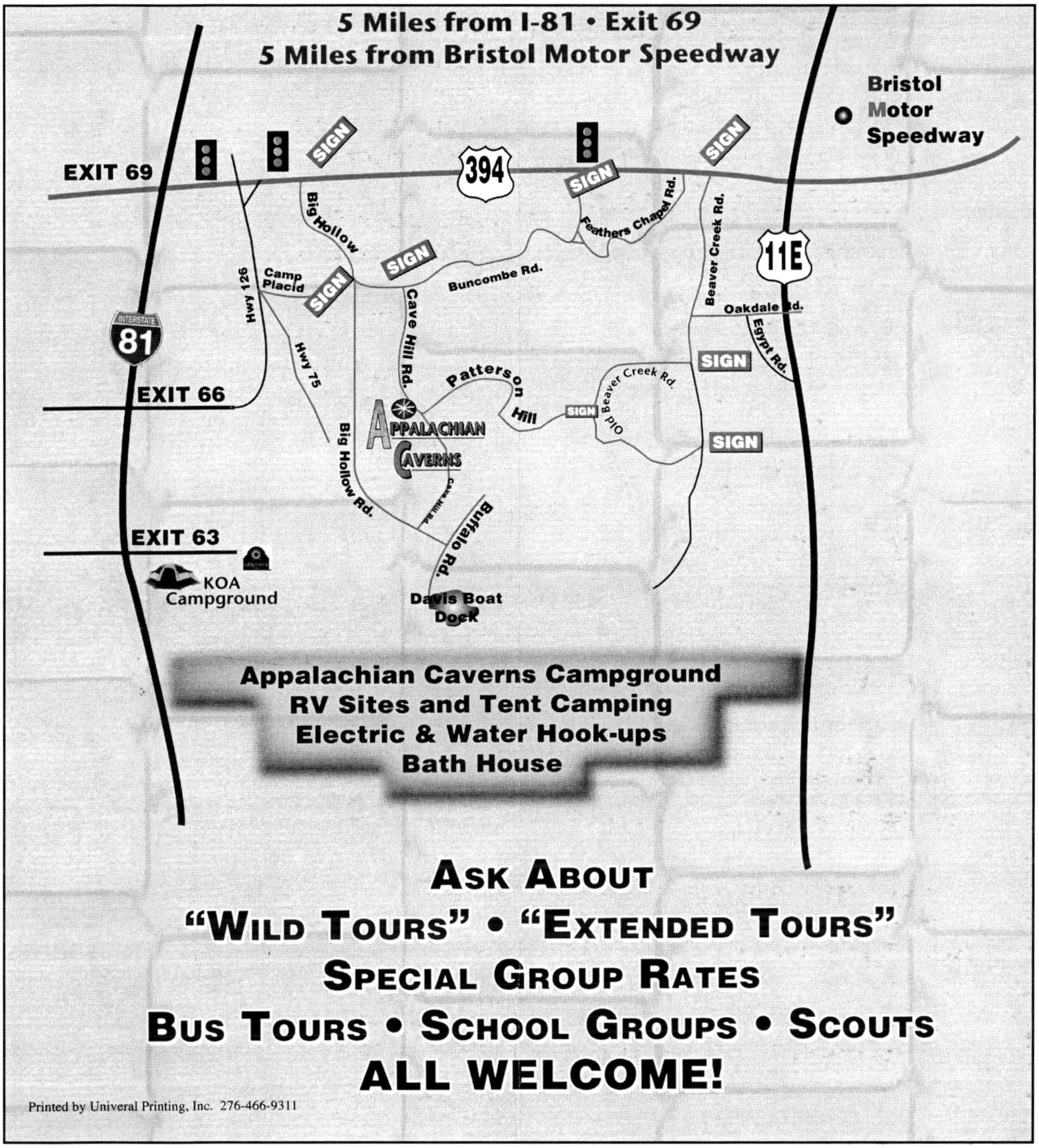

Figure 12.12. Road map inside of the current brochure for Appalachian Caverns. From the Larry E. Matthews collection.

highway that goes through Gray, Tennessee. In 2000, geologist Larry Bolt noticed that bones had been uncovered by a bulldozer. Subsequent investigations showed that a rich fossil find had been discovered. The bones date to the Miocene Epoch, which began 7,000,000 years ago and ended 4,500,000 years ago. The fossils uncovered to date include saber-toothed cat, tapir, rhinoceros, shovel-tusked elephant, and crocodile. For further information, visit the museum's Web site at:

http://www.grayfossilmuseum.com.

This museum is state of the art and well worth a visit. You can go outside into the excavation area and observe paleontologists excavating fossil bones. Also on display is a full-size replica of Sue, the largest *Tyranosaurus rex* skeleton that has ever been discovered.

Figure 12.13. Erica Sughrue on the arched bridge in Appalachian Caverns. Photo by Bob Biddix, March 23, 2008.

Figure 12.14. Erica Sughrue on the long bridge on commercial trail in Appalachian Caverns. Photo by Bob Biddix, March 23, 2008.

Figure 12.15. Sign advertising the Wild Tour in Appalachian Caverns. Photo by Larry E. Matthews, March 20, 2008.

Figure 12.16. Bob Biddix (L) Erica Sughrue (R) at a fork in the passage. Photo by Bob Biddix, March 23, 2008

Figure 12.17. Erica Sughrue explores a passage on the Wild Tour in Appalachian Caverns. Photo by Bob Biddix, March 23, 2008

Chapter 13

Bristol Caverns

The Discovery of Bristol Caverns

Bristol Caverns is located in Sullivan County, Tennessee. Although Tennessee did not officially become a state until 1796, Caucasian explorers had been visiting this area since at least 1743, when Dr Thomas Walker led an expedition as far west as present-day Kingsport. In the time period between 1743 and 1796, enough people migrated to Tennessee to successfully petition the United States government for admission as a state. Kentucky was the first state west of the Appalachian Mountains admitted to the Union and Tennessee was the second.

Due to the formidable mountains that separated Tennessee and Kentucky from Virginia and North Carolina, most of these early pioneers entered from Virginia, and the city of Bristol sits on the state line separating Virginia from Tennessee. New settlements then extended southwest, into what is now Tennessee, along the naturally occurring valley floors and the streams and rivers that occupy them. Other settlers moved through the Cumberland Gap into what is now Kentucky. The Indian tribes that had previously occupied this area were forced southwest, into what is now the Chattanooga area and further south into Alabama and Georgia.

According to legend, local Indians used the underground river as an attack and escape route in their raids on white settlers. The extreme ruggedness of this cave makes this story seem unlikely. The year that the cave was discovered by settlers is not known, but it could have occurred prior to 1800.

A Description of the Cave

Bristol Caverns is currently open to the public and is one of the premier show caves of east Tennessee. The book, *Caves of Tennessee*, which was written by Thomas C. Barr Jr, in 1961 gives the following written description of the cave:

> Bristol Caverns is one of the largest and most spectacular dripstone caves in Tennessee. The cave consists of a large gallery 2,500 feet long, which strikes north-northeast and averages 60 feet in width. The height is very irregular, varying between 10 feet and 90 feet because of breakdown covering the floor. A large stream, Sinking Creek, flows through the lowest level of the cave, winds back and forth through the main gallery, and emerges half a mile west of the entrance at the bottom of a bluff. Several attempts have been made to dam this stream to supply power for a grist mill, but the stream consistently has found another outlet.
>
> According to legend, Betsy Bishop, a young widow, was stabbed and thrown into the original entrance of the cave—a pit-like hole in the roof of the main gallery—by two brothers, hirelings of a jealous rival. Betsy's son is said to have returned years later and killed both brothers. For many years the cave was known as Bishops Cave. Over several decades the property surrounding the cave was acquired by A.H. Hitt, who opened part of the caverns to the public after World War II.
>
> The present entrance was discovered by digging out a spring and is only 30

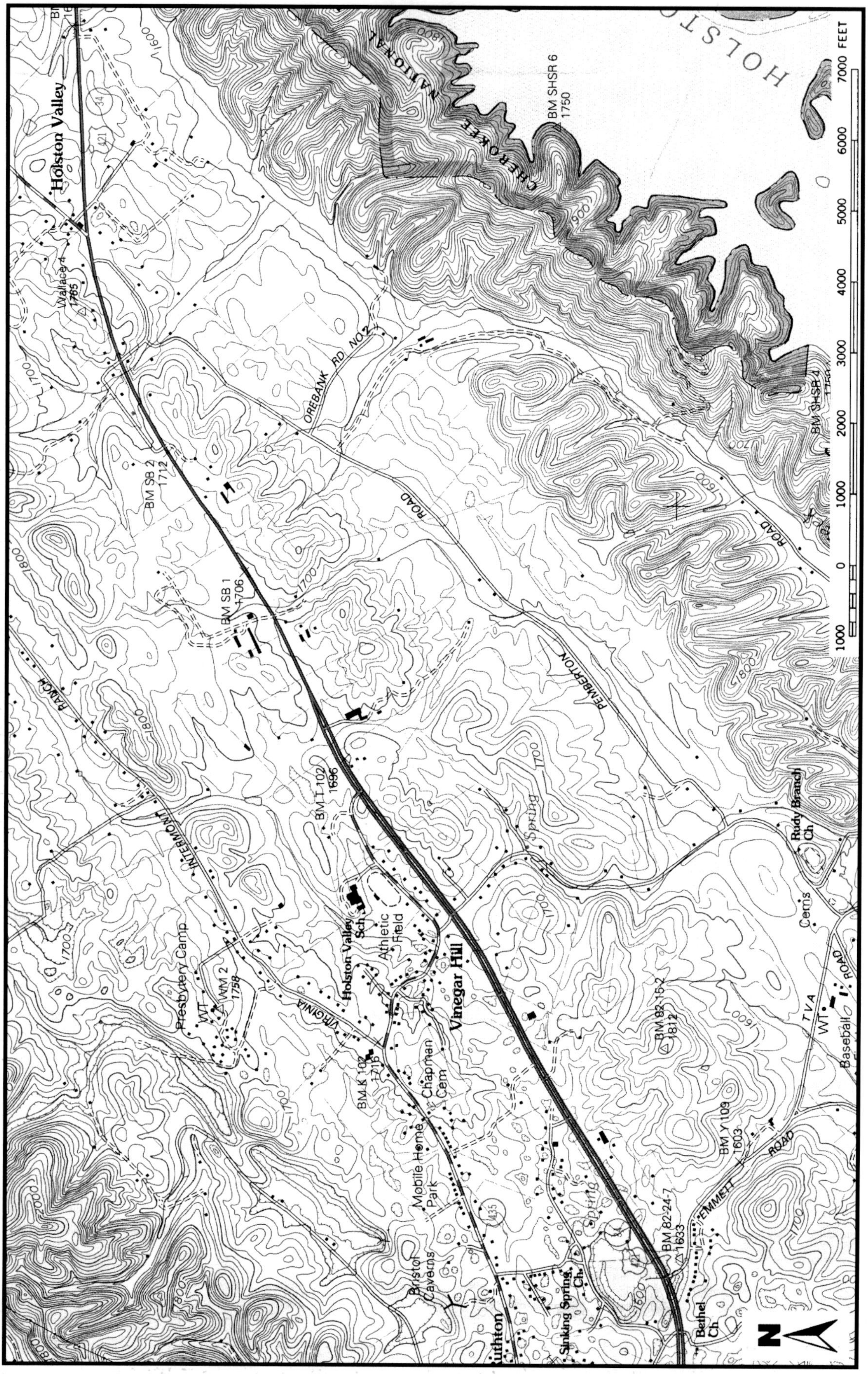

Figure 13.1. Topographic map showing the location of Bristol Caverns.

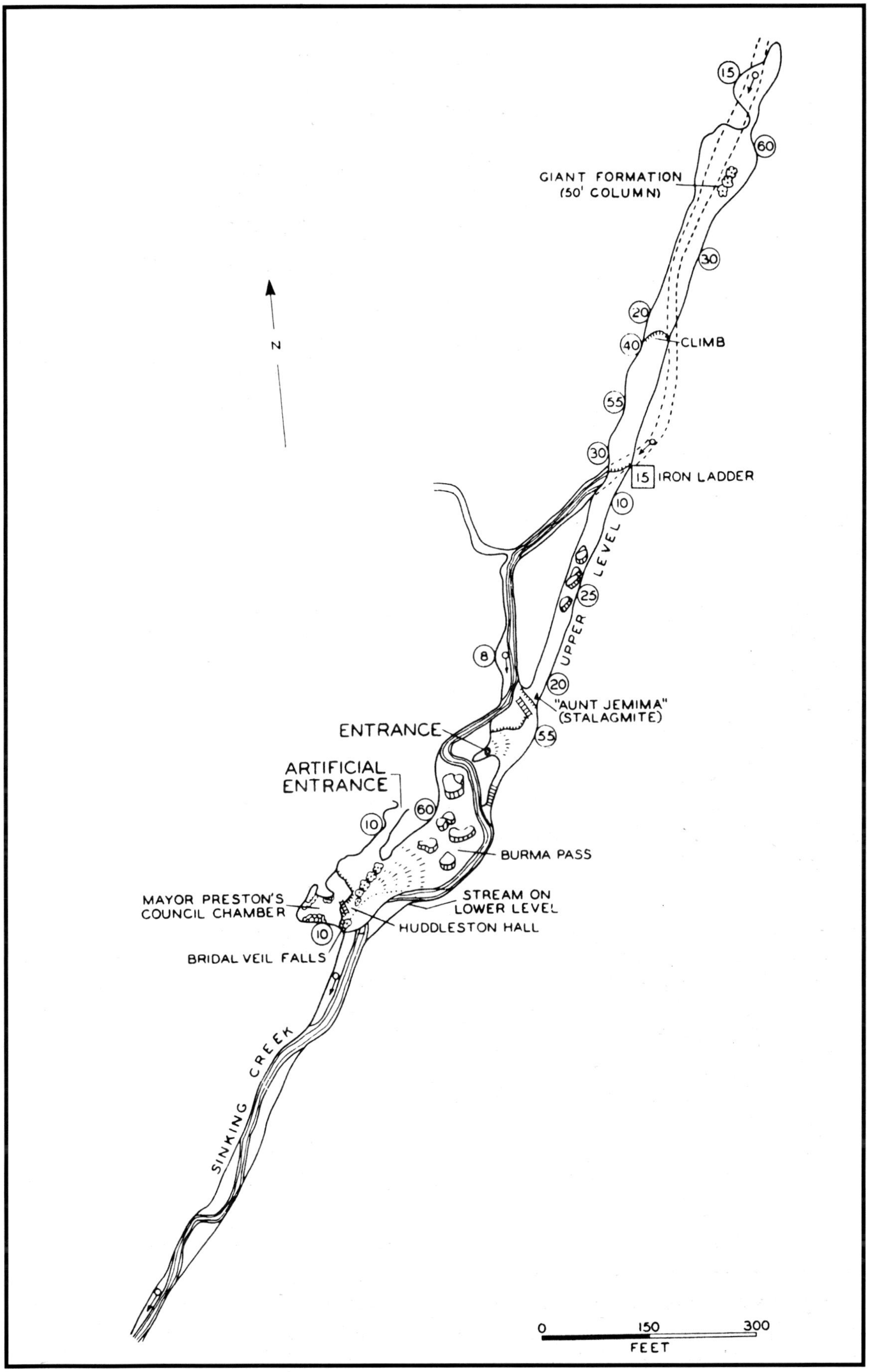

Figure 13.2. Map of Bristol Caverns from Barr's Caves of Tennessee, page 432.
Survey by A.H. Hitt and J.M. Kellberg (1942), with additions by T.C. Barr (1954).

Figure 13.3. Front and back of an old brochure for Bristol Caverns. From the Speleo Archives of Gary K. Soule.

feet from the original entrance, which is now closed by iron bars. The cave opens near the top of a low ridge at the head of a valley. At the southwest end of the cave one enters the upper part of the gallery, an area almost occluded by massive and beautiful dripstone columns and stalactites. Mayor Prestons Council Chamber (named for a former mayor of Bristol), Bridal Veil Falls, and Huddleston Hall are along the points of interest here. Beyond Huddleston Hall, a vast canyon known as Burma Pass,

Figure 13.4. Inside of an old brochure for Bristol Caverns. From the Speleo Archives of Gary K. Soule.

yawns below the visitor, with Sinking Creek flowing far below. Crossing the creek on a bridge, one enters a huge room with a "skylight" in the top—the natural entrance. Here the casual visitor or tourist must retrace his steps, but the explorer may continue.

A large stalagmite, named "Aunt Jemima," provides convenient footholds to clamber up around a ledge 50 feet above the creek. From there one can proceed into the remote portions of the cave. A 10-foot

Figure 13.5. Very old black and white postcard of Bristol Caverns showing the entrance to Mayor Prestons Conference Room. From the Gordon and Judy Smith collection at the National Cave Museum, Diamond Caverns, Park City, Ky.

Figure 13.6. Very old black and white postcard of Bristol Caverns showing the Jewel Room. From the Gordon and Judy Smith collection at the National Cave Museum, Diamond Caverns, Park City, Ky.

Figure 13.7. Very old black and white postcard of Bristol Caverns showing the Pipe Organ. From the Gordon and Judy Smith collection at the National Cave Museum, Diamond Caverns, Park City, Kentucky.

steel ladder provides a means of negotiating a short descent, and the explorer enters a vast hall of towering stalagmites. Beyond Aunt Jemima the stream is not in evidence, and apparently it flows far below. Beyond the stalagmite hall is the most awe-inspiring sight in the entire cave—a huge dripstone formation 52 feet high, 35 feet wide, and 10 feet thick. This is the largest known speleothem in any Tennessee cave and is composed of three fused columns, each of which alone is as large as the celebrated "Giant Dome" in Carlsbad Caverns, New Mexico. This tremendous mass of dripstone is known as the "Giant Formation."[1]

1 Thomas C. Barr Jr, *Caves of Tennessee*, Bulletin 64, Tennessee Division of Geology, 1961, pp 431–435.

Figure 13.8. Very old black and white postcard of Bristol Caverns showing an early ticket office. From the Gordon and Judy Smith collection at the National Cave Museum, Diamond Caverns, Park City, Kentucky.

Figure 13.9. Old postcard showing "Bridal Veil Falls" in Bristol Caverns. From the Speleo Archives of Gary K. Soule.

Figure 13.10. Old postcard showing "Hiddleston [sic] Hall" in Bristol Caverns. From the Speleo Archives of Gary K. Soule.

Figure 13.11. Old postcard showing the "Entrance to Mayor Preston's Conference Room" in Bristol Caverns. From the Speleo Archives of Gary K. Soule.

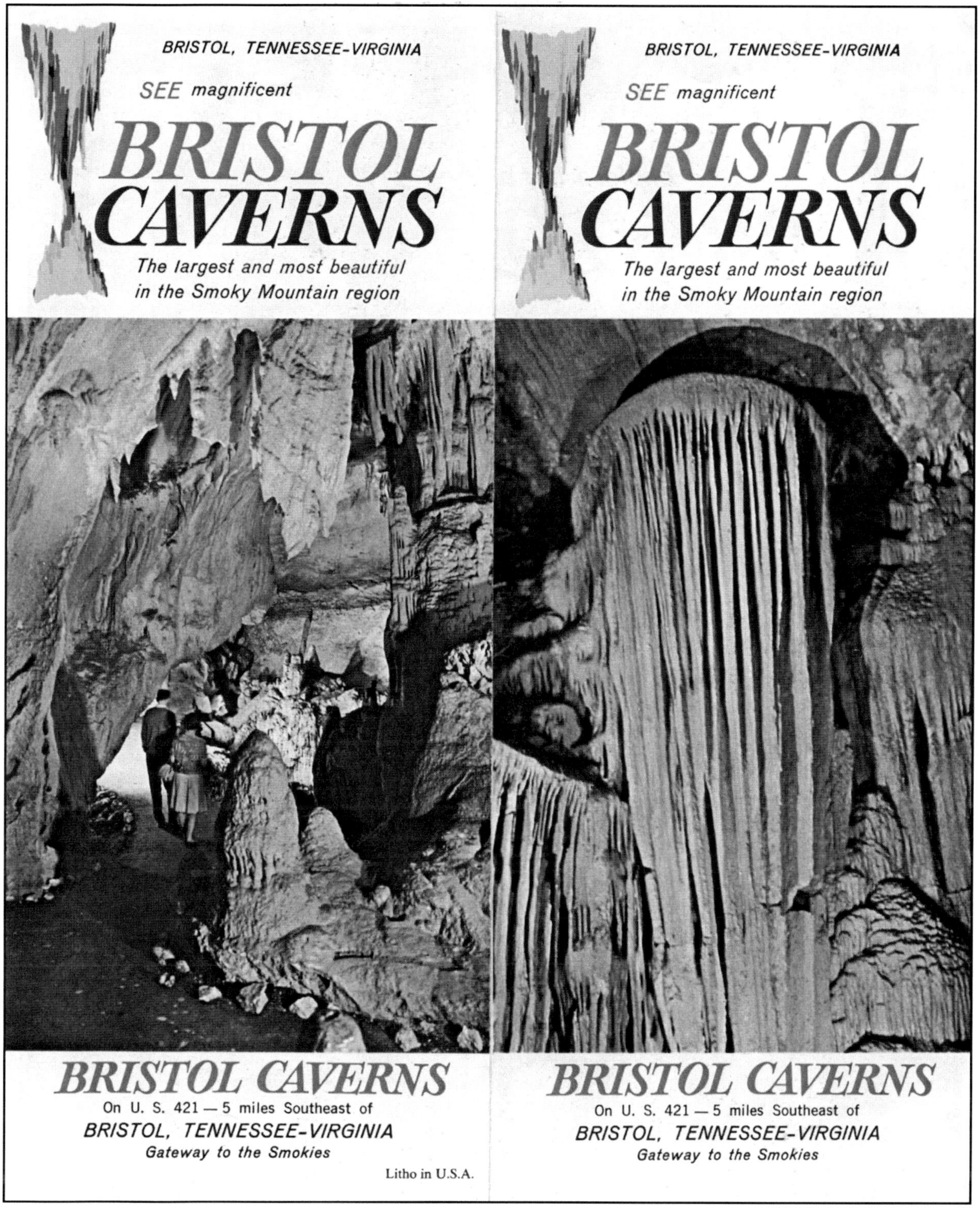

Figure 13.12. Front and back of a brochure for Bristol Caverns. From the Speleo Archives of Gary K. Soule.

The original entrance to Bristol Caverns was a small hole (the Skylight) that is now covered with iron bars for safety. The commercial entrance was dug into the hillside just a hundred feet away so that tourists could easily walk into the cave. This cave is extremely well decorated with speleothems. Another tunnel entrance was begun close to the current commercial entrance, but never finished. It would have provided an exit so that the tour would not have to retrace the path back out through the existing commercial entrance. As noted by Barr, there is a Stream Entrance half a mile west of the commercial entrance where

Figure 13.13. Postcard showing two women standing on a bridge over the cave stream in Bristol Caverns. From the Speleo Archives of Gary K. Soule.

the cave stream flows out at the base of a bluff.

Mr. A.H. Hitt purchased Bristol Caverns in 1933 and operated the cave until 1956. At that time, he was forced to sell the cave due to ill health.[2]

2 L.P. Foster, "Famed Bristol Caverns Go on Auction Block August 31," *NSS News*, vol 14 (1956), no. 9, p 86.

The Tennessee Long Caves List ranks Bristol Caverns as number 113 in length in Tennessee. The USA Long Cave List ranks Bristol Caverns as number 623 in length in the United States, with a length of 8,050 feet or 1.525 miles (2.454 kilometers).[3]

3 Bob Gulden, *USA Long Cave List*, April 16, 2007.

Figure 13.14. Postcard showing two women standing in a forest of giant cave formations. From the Speleo Archives of Gary K. Soule.

Figure 13.15. Front of a brochure from Bristol Caverns with the Interstate Highway 81 emblem. From the Speleo Archives of Gary K. Soule.

Bristol Caverns Today

Bristol Caverns is currently open to the public. Tours leave about every 20 minutes and last approximately 1 hour. The cave is open year round except for Easter Sunday, Thanksgiving, Christmas Eve, and Christmas Day

For more information on Bristol Caverns, you may telephone them at:

423-878-2011

The mailing address for Bristol Caverns is:

Bristol Caverns
PO Box 851
Bristol TN 37620-0851

Or, you may e-mail them at:

BristolCaverns@aol.com.

Bristol Caverns also has an informative Web site, located at:

http://www.bristolcaverns.com

Directions

From Knoxville, go east on I-40 for 34 miles to the I-81 junction. Go north on I-81 and proceed northeast for 69 miles to Exit 69. Go right (east) on Tennessee Highway 394. After about 6 miles you will pass the Bristol Motor Speedway. After 13.2 miles on Tennessee 394 you will come to an intersection with U.S. Highway 421. Turn right on U.S. Highway 421. After 4.5 miles, turn left (south) onto Bristol Caverns Road (Tennessee Highway 435). Proceed 1.5 miles and the visitor center for Bristol Caverns is on the right.

Side Trips

The world-famous Bristol Motor Speedway is located nearby on U.S. Highway 11. You may wish to schedule your visit so that you can tour Bristol Caverns and attend a race on the same trip. If so, be sure to make motel reservations well in advance.

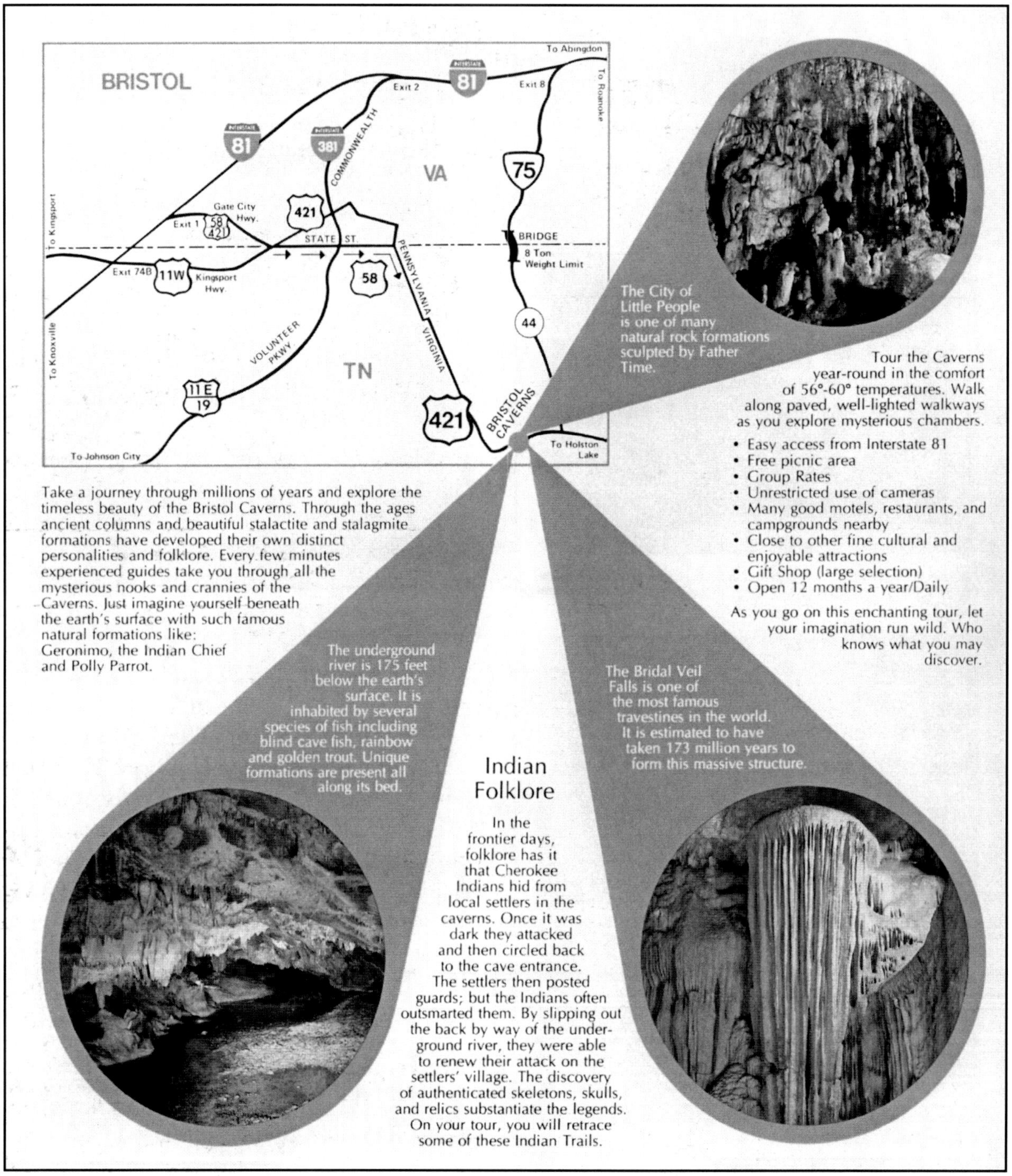

Figure 13.16. Inside pages of a recent brochure for Bristol Caverns. From the Speleo Archives of Gary K. Soule.

Figure 13.17. Betty Matthews in front of the visitor center for Bristol Caverns. Photo by Larry E. Matthews, March 19, 2008.

Figure 13.18. The Bristol Caverns sign along Bristol Caverns Road. Photo by Larry E. Matthews, March 19, 2008.

Figure 13.19. The Historic "Skylight" Entrance to Bristol Caverns. Note soft drink bottle for scale. Photo by Larry E. Matthews, March 20, 2008.

Figure 13.20. The Commercial Entrance to Bristol Caverns. Photo by Larry E. Matthews, March 20, 2008.

Chapter 14

Morrill Cave

The Early History of Morrill Cave

Just inside the upper entrance to Morrill Cave there was a prehistoric stone wall some 6 feet high, built of rocks of various sizes, which is considered by some to be evidence that aboriginal people inhabited the cave. Archaeologist Gerald Fowke excavated this wall in the early 1900s. However, he found no associated artifacts with this wall that would indicate a probable date for its construction.

Sullivan County was formed in 1779, so clearly settlers were living in this area before that time. However, nothing is known of the cave until it became the property of Elias S. Worley sometime before 1850. Locally, the cavern is often still referred to as Worleys Cave.

As noted above, the first owner of the cave was Elias S. Worley. The exact date he acquired the land is unknown, but he is listed in the 1850 Census. Cave historian Laurie Adams gives the following information about Worley:

> Though E.S. Worley owned the land, it is uncertain what degree of involvement he had with the cave. His signature, the most distinctive and famous in the cave, is located in the breakdown room just beyond the "skating rink" and the tombstone area. In large, old-style script, enameled onto a breakdown block was a black pigment inscription: "E.S. Worley – J.R. Cunningham – 1884 – 'Giant's Causeway." His signature has not been found elsewhere in the cave, and since the Signature Room, as it is now known, is the farthest point penetrable into the cave without requiring crawling or wading, perhaps this is as far as he penetrated.[1]

A large amount of saltpeter was mined from the cave early in the Civil War and that will be discussed in more detail later in this chapter. Another commercial use of the cave was a mill that was operated in the early 1900s where the stream exits at the lower entrance of the cave.

A local resident, John Morrill, led many explorations of the cavern "near the turn of the century." Like many caves, this cave has gone by a variety of names over the years. It has been called Morrell Cave, Morrill Cave, Morrils Cave, Morrels Cave, Mirrells Cave, Worleys Cave, and Worlies Cave. In 1980, the U.S. Board on Geographic Names attempted to resolve this issue when it approved the name Morrell Cave. Clearly, the U.S. Board on Geographic Names is in error in this case. The Tennessee Division of Geology had officially described this cave as Morrill Cave in 1961 in Thomas C. Barr Jr's book *Caves of Tennessee.* This name precedes the decision of the U.S. Board on Geographic Names and should, therefore, take precedent. In this book, it shall be called Morrill Cave.

Saltpeter Mining

Cave historian Laurie Adams researched the history of saltpeter mining in Morrill Cave. Here are his findings:

> This brings us to the question of

1 Laurie Adams, *History of Morrell Cave, Part I*, Published August 1984 by the Flittermouse Grotto of the NSS, 11 pages.

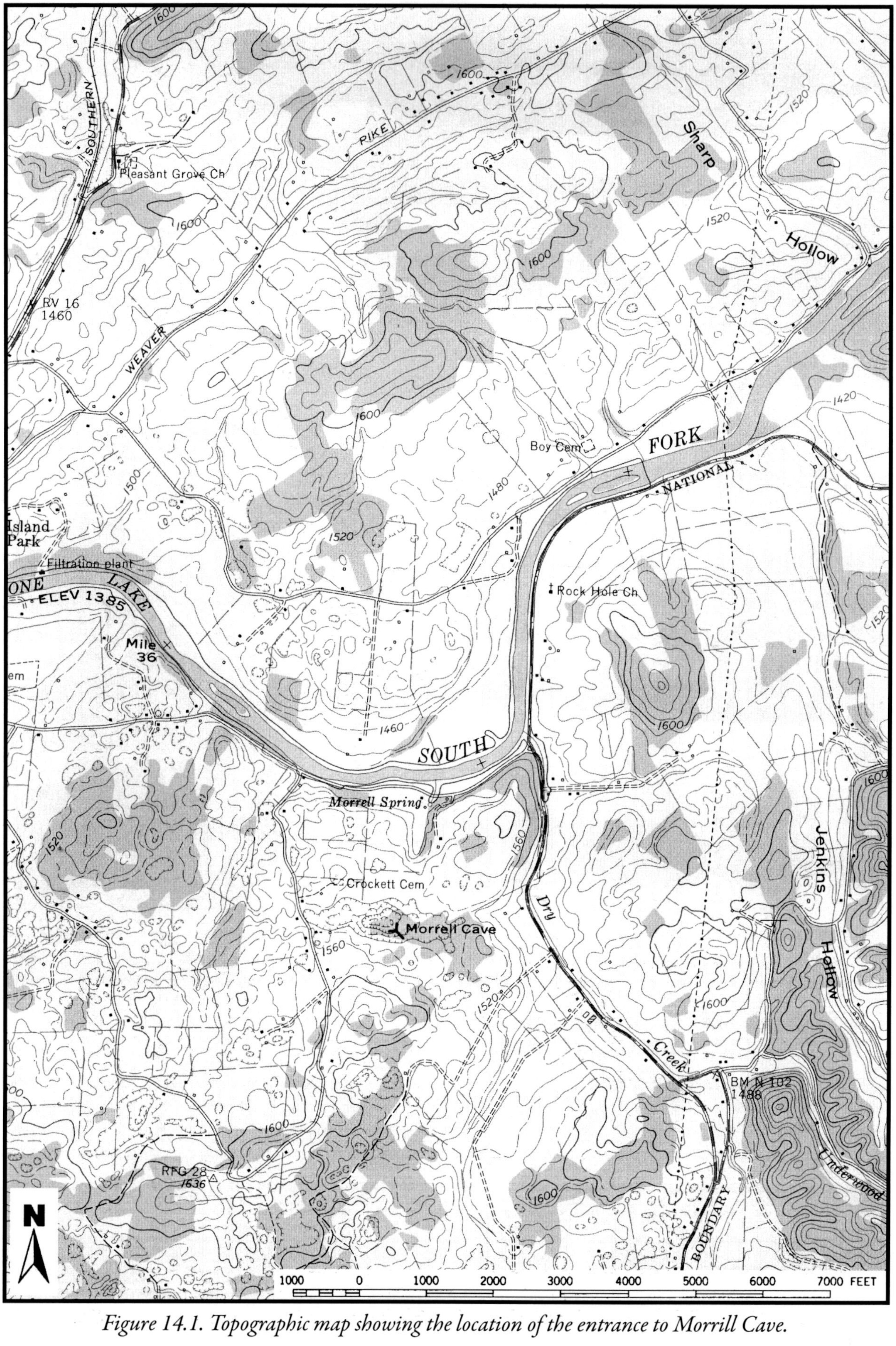

Figure 14.1. Topographic map showing the location of the entrance to Morrill Cave.

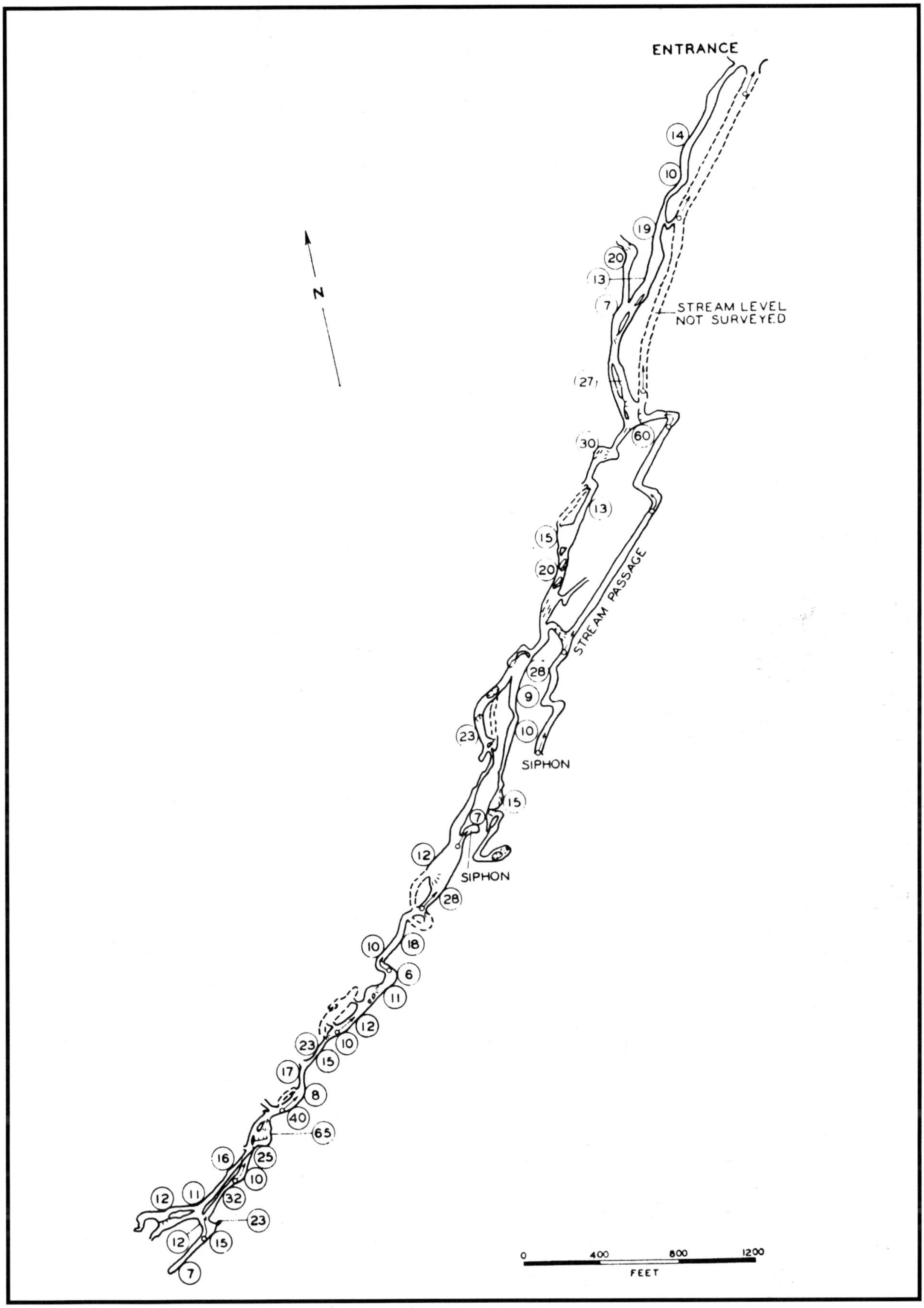

Figure 14.2. Map of Morrill Cave from Barr's Caves of Tennessee, *page 437. Survey by L.C. Johnson, L.Powell Foster, G. Kirk Finch, and John Tamblyn, 1955–1956.*

saltpeter mining, and there are certainly more questions than answers. Did Worley own the cave during the Civil War when it was mined? I believe he did, for this reason. There are only three sources of information about saltpeter mining in the cave of which I am aware. The first is the *Compiled Service Records, Record Group 109, National Archives – Detailed Conscripts of the Sullivan County Nitre and Mining Bureau.* The second is Gerald Fowke's account, already quoted. We know that Worley owned the cave at the time of Fowke's excavation, sometime near the beginning of the (20th) century. Where else would Fowke have gained his information about the miners, except from Worley himself? The fact that he (Worley) was on the 1850 census indicates that he was of sufficient age by the 1861–1865 wartime period to have been aware of the mining activity taking place on his or perhaps his father's land, so it is probably him we have to thank for passing on his knowledge. The third reference is one sentence in L.P. Foster's caption on page 66 of the July 1956 *NSS News*, where he simply states, "The upper level of the cave was mined for saltpeter during the Civil War, and the nitre-containing dirt was floated out of the cave in boats on the stream in the lower level." Perhaps foster obtained his information from Fowke's account.

The only clue as to the date of mining is Fowke's statement...*"Early in the Civil War ..."* This would indicate a date of 1861 to maybe 1863. But who were these elusive miners? We don't know for sure; the record is very incomplete and there are no direct statements that certain people worked in Morrell Cave. There are some hints

Figure 14.3. Erica Sughrue at the Stream Entrance to Morrill Cave. Photo by Bob Biddix, March 22, 2008.

Figure 14.4. Old names and dates, including J.N. Johnson, March 26th, 1882, in Morrill Cave. Photo by Sandra S. Sherlin, January 1, 2006.

Figure 14.5. Old names and dates, including N.L. Davis, May 17th, 1881, in Morrill Cave. Photo by Sandra S. Sherlin, January 1, 2006.

Figure 14.6. Erica Sughrue at the Upper Entrance to Morrill Cave. Photo by Bob Biddix, March 22, 2008.

*Figure 14.7. Old name and date—D.G. Eads, 1858—in Morrill Cave.
Photo by Sandra S. Sherlin, January 1, 2006.*

Figure 14.8. Erica Sughrue looking out the Upper Entrance to Morrill Cave. Photo by Bob Biddix, March 22, 2008.

Figure 14.9. Pick marks left in the dirt by saltpeter miners during the Civil War. Photo by Bob Biddix, March 22, 2008.

Figure 14.10. Pine torches believed to have been left by saltpeter miners during the Civil War. Photo by Bob Biddix, March 22, 2008.

in the *Detailed Conscripts of the Sullivan County Nitre and Mining Bureau.* (I am indebted to Marion O. Smith of Knoxville, Tennessee, for uncovering the following information.)

Two companies were organized in April, 1864, composed of conscripts for local defense in Sullivan County. They were also detailed as laborers in the Nitre and Mining Bureau, District No. 7, which was under the command of Captain (later Major) Thomas James Finne. John W. Pearce was Captain of Company A. B.W.L. Holt was Captain of Company B. By cross-referencing names from the rosters of Companies A and B with names of men making application for a Tennessee Confederate Pension, the following data was obtained.

Sergeant J.I. Harkleroad – Company B

Described during the war as age 25, 5 feet 10 inches tall, fair complexion, blue eyes, light hair. Farmer, 16th District, Sullivan County, Tenn. Confederate Pension Application No. 7908. J.I. Harkleroad – born in Sullivan Co., 1837. A 1906 statement of B.H. Richards: *"that he [Richards] knows Mr. Harkleroad was in Maj. Finney's Mining and Nitre Bureau. That in 1862, he saw him [Harkleroad] performing this labor near Bluff City, and in 1863, he [Harkleroad] went with affiant [Richards] and others to Va. in same service and remained up there with us, about 5 weeks ... he [Harkleroad] then returned to Bluff City, Tenn. and resumed the same work there again ... [and] was in this work near Bluff City, Tenn. at the time Gen. Lee surrendered."*

Private W.H. Carrier – Company B

Described in April, 1864, as age 27, 5

Figure 14.11. Erica Sughrue in a cave passage in Morrill Cave. Photo by Bob Biddix, March 22, 2008.

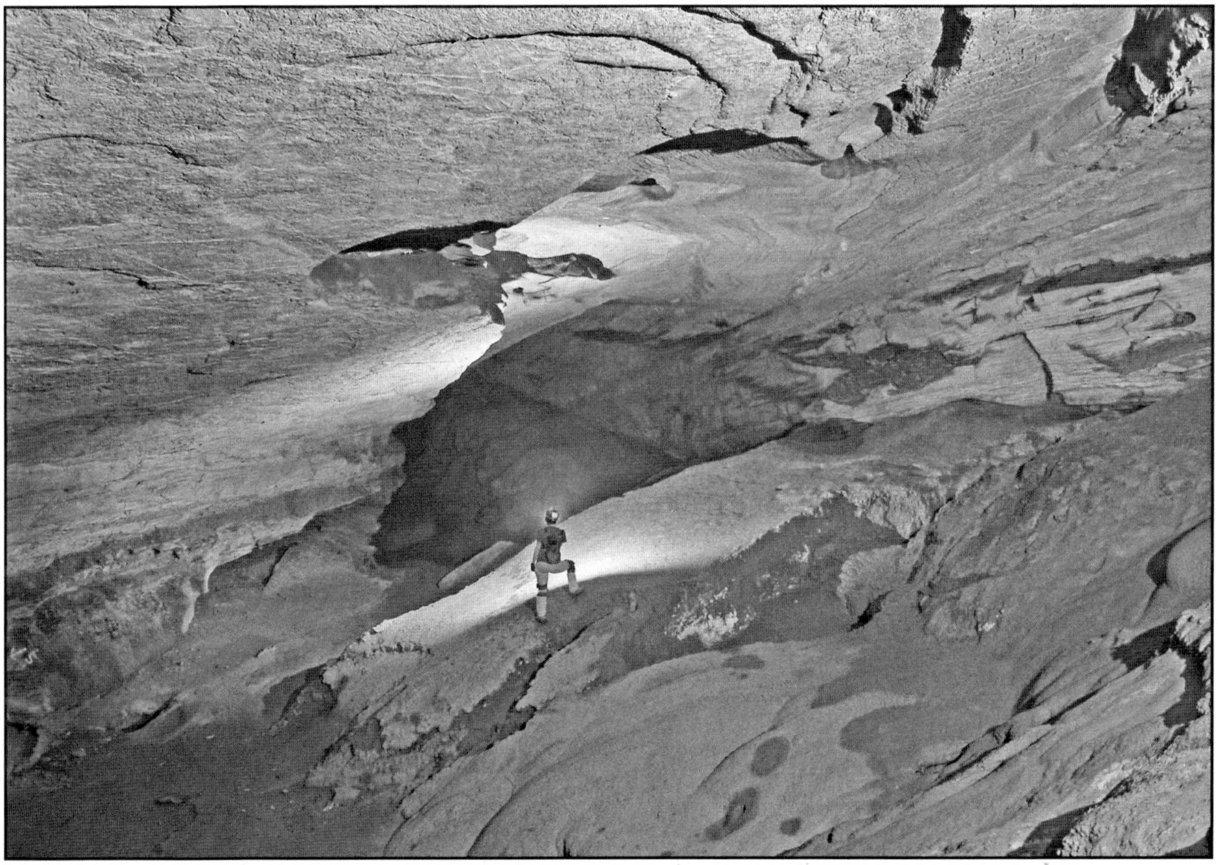

Figure 14.12. Erica Sughrue in a large passage in Morrill Cave. Photo by Bob Biddix, March 22, 2008.

feet 8 inches tall, fair complexion, blue eyes, light hair. Farmer and Cooper. Confederate Pension Application No. 8594. W.H. Carrier – born in Sull. Co. in 1836. He made the statements: *"I was detailed to make barrels for the Nitre works. I worked most of the time in Bluff City. After General Robert E. Lee surrendered, I was just released by Capt. Pearce, at Bluff City and told to go home."*

Figure 14.13. Erica Sughrue under a cantilevered ceiling in a large passage in Morrill Cave. Photo by Bob Biddix, March 22, 2008.

Private John K. Ensor – Company B

Described during the war as age 33, 5 feet 10 inches tall, fair complexion, grey eyes, light hair. Farmer, 18th District, Sull. County, Tenn. Confederate Pension Application No. 8148. J.K. Ensor – born in Wash. Co., Tenn. 1832. On April 3, 1906 C.A. Hodges stated: "*That he [Hodges] was a member of same command with the applicant and worked side by side with him .... When the war closed he [Ensor] was still at work in the nitre mine in Sullivan County, and after it was all over and General Lee surrendered, the applicant [Ensor] and affiant [Hodges] just stopped their work for the Confederate Government.*"

From these and other pension applications it is clear that the Nitre and Mining Bureau was engaged in saltpeter mining in Sullivan County during the war. There is no written statement that they worked at Morrell Cave.

The physical remains of mining activity in Morrell Cave can tell us a little. The most prominent evidence is the extensive trail system developed through the upper level as far back as the beginning of the "Onward Crawl." There is a definite pattern to the trails. They are absent in the main upper level from the entrance as far back as the 2nd stream connector, located just beyond a split-off rock, about 1,000 feet

Figure 14.14. Erica Sughrue by a large flowstone formation in Morrill Cave. Photo by Bob Biddix, March 22, 2008.

Badly rusted hand-wrought iron nails have been found in the room. I believe if the site were properly excavated, it might reveal further information. From the Saltpeter Room back into the cave, there is only a patchy trail system, cuts being made only to eliminate the most awkward spots.

The layout of these trails sheds light on the statements made by Gerald Fowke, who stated: "... A dam was constructed just within the mouth of the main cave, and in the pool thus formed boats were used to transport the material from the interior. The workmen not required for handling the craft usually preferred to walk through the upper cave to the place where the earth was poured."

into the cave. From here a well-developed trail extends back to the "Saltpeter Room," located just before the beginning of the Railroad Passage. The Saltpeter Room is a section of passage about 130 feet long and averaging 25 feet wide. An enormous amount of dirt was removed from this area, much more so than any other site in the cave. There are several cuts [work faces] still showing well-preserved pick marks, piles of saltpeter rocks, torch marks, piles of leached dirt, and even remains of vats or some other type of wooden structure.

We now know of the existence of a sump located about 1,500 feet from the entrance, (The Cox Sump) and because of it, they could not have brought the nitre-bearing dirt from very far into the cave using only a water route. The problem is cleared up by the existence of the 2nd stream connector, where they could have brought the dirt down from the upper level and loaded it on a boat. This would explain why there is no trail system in the upper level from the

Figure 14.15. Erica Sughrue admires flowstone formations on the wall in Morrill Cave. Photo by Bob Biddix, March 22, 2008.

entrance to the 2nd connector, yet from this point back there is a good trail system. In addition, we know that the upper entrance was a crawlway before being dug out by Fowke in the early 1900s. This also points to the likelihood of a second dam constructed in the Big Room, about half a mile into the cave, where the stream level and upper levels once again intersect. Had they built a dam here they would have been able to bring dirt out from way back in the cave, utilizing an upper to lower level connector situated a couple of thousand feet farther back from the Big Room. It would have been impractical if not impossible to use the upper level from the Big Room back, as there are many obstacles: climbing over breakdown, skirting pools, negotiating climbs, etc. As further evidence to support this idea, there mysteriously appears a good trail system beginning at this connector and extending farther back into the cave.

With a two-dam system, they would have been able to bring dirt from as far as the Onward Crawl, carry it to the stream connector, load it onto boats and float it down to the Big Room. Here they would have once again carried it through the upper level to the 2nd stream connector, where they would once again load it on boats and float it to the entrance. This undoubtedly would have been a lot of work, and that is probably why there was not a great deal of dirt removed from deep in the cave. The Saltpeter Room would have been the easiest source of dirt to procure in the entire cave. Unfortunately, no recognizable remains of either dam are present today.[2]

2 Ibid.

Figure 14.16. Erica Sughrue by a large rimstone dam in Morrill Cave. Photo by Bob Biddix, March 22, 2008.

An Archaeologist Visits Morrill Cave

Gerald Fowke was a researcher with the Bureau of American Ethnology who traveled throughout the United State conducting archaeological digs. Most of his work was done in caves, particularly in Missouri and Tennessee. In 1922 he published a book titled *Archaeological Investigations.* Below is his description of Morrill Cave:

> Morrell Cave. On the south side of the Holston River, 2½ miles east of Bluff City, lies the farm of E.S. Worley. Except for a narrow strip of river bottomland, the surface is broken and rocky, the highest point being some 400 feet above the stream. Beginning near the brow of the river hill, the central portion of the farm is in a depression whose very irregular rim or watershed surrounds an area of more than 100 acres. All the water that falls within this space drains into a sinkhole, the bottom of which is but little above flood stage of the (South Fork) Holston. On the south side of this sink is a vertical bluff 120 feet high, from whose foot emerges a stream that, after a winding course of 50 or 60 yards, disappears in a small opening on the east side of the sink hole and finally comes to the surface at the foot of the hill, near the river. Its volume is sufficient, even in times of severest drought, to turn the undershot wheel of a large mill. The course of the stream above the point where it is first visible is through a cave which has been traced to the foot of the Holston Mountains 3 miles away, and there are many unexplored branches. Chambers are known with a cross measure of 100 feet or more, and some of them have a height nearly as great. Stalactites and stalagmites, some of them possessing unusual size and beauty, are abundant. The sinkhole is due to the falling in of the roof of the cave, which no doubt could

be followed to the river if it were free from obstructions in this direction.

North of west from the mouth of the cave is another opening, partly in the same strata but 40 feet higher, the dip of the rock being 10 or 12 degrees to the southeast. This was so blocked with talus which had fallen from the cliff and washed down the side of the sinkhole that it was necessary to creep nearly 40 feet from the entrance, down a moderate slope, before coming to a point where it was possible to stand upright. From here progress to the junction of the two caves, about half a mile from the entrance, is easy except where fallen rocks interfere somewhat.

Figure 14.17. Erica Sughrue at a huge flowstone formation in Morrill Cave. Photo by Bob Biddix, March 22, 2008.

Early in the Civil War a large amount of saltpeter was manufactured here. A dam was constructed just within the mouth of the main cave, and in the pool thus formed; boats were used to transport the material from the interior. The workmen not required for handling the craft usually preferred to walk through the upper cave to the place where the earth was procured.

The combination of natural features at this place is unusually favorable to aboriginal habitation. The main cave is excluded from consideration by reason of the stream filling it from wall to wall after very heavy rains. The upper cave, however, showed, beyond the debris chocking the entrance, a level floor, cumbered, it is true, by fallen rocks, but apparently quite suitable for a dwelling place were these removed. Although opening toward the north, its position so far below the summits of the surrounding hills protects it from winter winds. The creek assures an ample supply

Figure 14.18. Erica Sughrue examines the giant Frozen Waterfall formation in Morrill Cave. Photo by Bob Biddix, March 22, 2008.

of clear, cold water. Mountains, refuge for game, are in sight in various directions, while the (South Fork) Holston River is less than a quarter of a mile away.

In order to remove the debris, a point 3 feet below the lowest spot on the floor was selected on the slope outside. From here a trench was carried in on a level, the additional depth being taken to facilitate clearing away all material that had accumulated inside the cavern in comparatively recent time, and thus lighten the task of deeper excavations should these be required.

The trench needed to be only wide enough at the bottom to allow room for running a wheelbarrow, but owing to the great amount of broken rock, loosely held together by a small quantity of earth, the sides continually gave way, so that by the time it was safe to pass through, the trench was 25 feet wide at the top and 24 feet deep at the mouth of the cave. The rocks were of every size from pebbles to blocks weighing more than a ton each.

Nothing whatever of artificial character, not even a flint chip or fragment of charcoal, was unearthed until at a point 4 feet inside the farthest projecting stratum of the roof. Here was found a prehistoric stone wall whose outer side and top had been entirely concealed by debris. On the inner side the upper portion was visible, owing to the fact that the owner had gathered a quantity of loose stones to construct a wall farther down the slope. Previous to this the ancient wall was entirely covered by the detritus, and even after this partial exposure its true nature was not suspected. It was about 6 feet high, built up of rocks of various sizes and shapes loosely fitted together, earth from the outside surface being used to level up in places where the stones would not bind properly. The largest rock in the top layer weighted [*sic*] about 800 pounds.

The horizontal distance between the top of the wall as it was when cleared off and the corresponding portion of the cave roof was 4 feet, to the roof directly above it, about 2 feet. Apparently it had at one time entirely closed the entrance; at the western end where it abutted against the solid rock the upper portion was firmly consolidated by travertine. Directly above it, nearly 2 feet higher, a slab and some small, irregular fragments were securely attached to the side and roof by the same agency. A crevice in the bedrock, just at the end of the artificial wall, contained several wagonloads of small rocks which had been thrown into it. These also were united into a solid mass by the travertine, all of which had been deposited by water flowing through the crevice. It does not follow that the wall was ever higher toward the opposite end than at this time. In the centuries that have elapsed since it was put up, the roof at the front of the cave, being rather thin-bedded, may have disintegrated. It was not possible to uncover the wall in shape for illustrating; portions of it continually crumbled as the looser material piled against it was removed.

From the wall inward the foreign material piled against the west side of the cave was composed almost entirely of small rocks, with scarcely any earth, and so compactly bound with travertine and stalagmite as to resist all attempts to remove it by ordinary means. On the east side—the left as the cave is entered—there was a great variation on the size of the stones; they were intermixed with much loose dry earth, and there was scarcely any "drip-formations" in the mass. The

Figure 14.19. Erica Sughrue on breakdown in a large passage. Photo by Bob Biddix, March 22, 2008.

removal of all this disclosed a projection of solid rock forming a shelf from 8 to 12 feet wide, whose top was about 2 feet higher than the bottom of our trench. About 20 feet from the ancient wall the trench reached the original bottom of the cave as the latter was left by the stream to which its origin was due. This was the tough red or yellow clay, filled with water-worn stones such as appear in all gullies or ravines in this region. It contained a small quantity of stalagmitic material here and there and gradually rose until at 20 feet farther, or 40 feet from the old wall, it terminated against solid bedrock, reaching across the cave, the entire width of which at this point was 26 feet. The shelf on the left belonged to the same stratum.

This brought the work to the terminus that had been the aim from the first, namely, the lowest level of the floor, which was thus shown to be only a foot above the solid rock instead of at least 10 or 12 feet, as the general appearance of the entrance and its surroundings had indicated. It was completely cleaned off as far as this was possible, but within 3 feet of the end of the trench began a mass several feet in thickness of fragmentary rocks of every size up to 20 tons or more which had fallen from the roof and were bound together by stalagmites.

Altogether, more than 300 cubic yards of material were removed. The workmen had been carefully instructed as to what the search was for, and kept a close lookout, as evidenced by the very small objects they were continually offering for inspection. It is safe to say that not a spadeful of earth missed scrutiny; but, aside from the artificial wall, the only traces of human presence were three valves of mussels, a

Figure 14.20. Erica Sughrue among dry rimstone dams in Morrill Cave. Photo by Bob Biddix, March 22, 2008.

> turkey bone rudely pointed for use as a perforator, and three or four bones which seem to have been subjected to fire. Not a chip of flint or other stone showing work, no ashes or charcoal, not a piece of pottery, were discovered. If aboriginal burials were made in the cave—and the wall is almost definite proof of such fact— they are either on the floor under stalagmite or in crevices now concealed by fallen rocks.
>
> Numerous small fragments of animal bones were found scattered singly at all depths in the material removed. Nearly every one showed marks of the teeth of rodents. According to Prof. F.A. Lucas, of the National Museum, they all belong to modern species except one tooth, which is that of the cave tapir, and (possibly) the jaw of an otter.

It seems amazing that Fowke excavated 300 cubic yards of material, and found nothing besides an enigmatic stone wall. If prehistoric peoples really built the wall, you would expect some artifacts to be present in the surrounding soil. For no artifacts to be present, is even more mysterious than the wall, itself.

A Description of the Cave

The book, *Caves of Tennessee*, which was written by Thomas C. Barr Jr, in 1961, gives the following written description of the cave:

> Morrill (or Worley) Cave is a very large and well-known cavern which opens in a huge sink 3 or 4 acres in area. It is the longest known cave in east Tennessee. There are two entrances, an upper dry mouth and a lower wet mouth, corresponding to the two levels of the cave, both located on the southeast side of the sink. The cave stream flows from the lower mouth, across the bottom of the sink, and disappears into the ground at the northeast side. It may well be the same stream that emerges from a low, cave-like opening on the bank of the Holston River at Morrils Spring, 0.4 mile away. A strong blast of air issues from this spring during warm weather.
>
> The wet mouth is 25 feet wide and 15 feet high and opens into a huge room fully 100 feet wide and 50 feet high. The dry mouth is 12 feet high and 18 feet wide. The cave is an excellent example of the Appalachian pattern, the two levels following the strike with but minor deviations, and the lower of the two developed down dip from the upper. For 2,000 feet the upper level is dry or damp, and earth floored, averaging 8 feet high and 15 feet wide.
>
> At 2,000 feet occurs the first junction of the upper and lower (stream) levels; here a large chamber is developed, 75 feet wide, 100 feet high, and 250 feet long. It contains two large stalagmites and many smaller formations, including a series of rimstone pools. The cave stream flows through the chamber and disappears under a rock ledge at the northeast end. Upstream the cave averages 25 feet wide and 15 feet high. It runs for an additional 2,200 feet to a point where the ceiling meets the water level.
>
> In the large room where the two levels meet, a climb extends up into the upper level, which may be followed to a point a mile and a half from the entrance. Just past this climb is a chamber containing a huge flowstone drapery 35 feet wide and 15 feet high. At several places the upper and lower levels intersect.
>
> The accompanying map of Morrill Cave has been prepared by L. Powell Foster and L. Clarke Johnson, of Kingsport, who have conducted an extensive study of the cave and of neighboring karst areas. East of the cave lies a limestone area riddled with caves, pits, and plugged sinkholes. Fully 25 of these solutional features have been charted by Foster and Johnson in the Walnut Grove area, in the range Latitude 36°25′19″ to 36°27′49″, Longitude 82°12′30″ to 82°15′00″, which lies between Holston River and the west base of Holston Mountain. Somewhere in this area is the source of the Morrill Cave

Figure 14.21. Erica Sughrue examines a cut flowstone formation in the Tombstone Factory. Photo by Bob Biddix, March 22, 2008.

stream, and it may well be from the many small tributaries arising in the solutional cavities of the Walnut Grove area. The water table in these caves is at 1,480 to 1,500 feet above sea level, 20 to 40 feet higher than the elevation of the sink at the mouth of Morrill Cave to the west. These karst features are located on the Keenburg Quadrangle.

The cave received its name from John Morrill, a local resident, who died in 1956. He led many cave explorations near the turn of the century. A more popular name at the present is "Worley Cave."[3]

The 2007 USA Long Cave List ranks Morrill Cave as #177 in length in the United States, with a length of 23,000 feet, or 4.4 miles (7.0 kilometers).[4] This makes Morrill Cave the second longest known cave in east Tennessee. Only Cuyler Cave in Sevier County is longer, with a length of 43,085 feet, or 8.2 miles (13.2 kilometers).

The Tombstone Factory

Morrill Cave has the unique distinction of being the only cave in Tennessee, and perhaps the entire United States, that was used as a tombstone factory. According to cave historian Jim Whidby, large stalagmites were "sliced up" and manufactured into tombstones and sold locally. Several tombstones in a nearby graveyard still exist that came from Morrill Cave.

Stalagmites are composed primarily of calcite, the same mineral that forms marble. Marble has been a popular stone for tombstones because of its attractive white color. The persons

---

3 Thomas C. Barr Jr, *Caves of Tennessee*, Bulletin 64, Tennessee Division of Geology, 1961, pp 436-440.

4 Bob Gulden, USA Long Cave List, April 16, 2007.

who manufactured tombstones also preferred marble because it is relatively soft and easy to carve. Unfortunately, the same solubility that allows caves to form in limestone (which is also composed of calcite) means that these tombstones weather fairly rapidly in our moist Tennessee climate. A visit to a local graveyard will show that many tombstones manufactured from marble are barely legible after only 100 years. Tombstones manufactured from flowstone stalagmites have suffered the same fate.

Figure 14.22. Erica studies another cut flowstone formation in the Tombstone Factory. Photo by Bob Biddix, March 22, 2008.

Cave historian Laurie Adams visited this site in the cave, but was unaware what industry had occurred there, even though he did know that this area of the cave was named the "tombstone area":

> There is one other mystery connected with this period in the cave's history. There are three sites in the cave where inexplicable slices were made into various forms of flowstone. Stalagmites, columns, and flowstone slopes were sliced by some unknown type of saw, with smooth even faces and with keen accuracy. The larger of the sites is known as the tombstone area and contains many cuts. The other two sites are in the vicinity of the Crystal Palace (or Fairies Ballroom) and they have only a few cuts. The fascinating aspect of these cuts is their age. There are dozens of old names and dates written on the smooth faces, mostly in pencil or charcoal. Unfortunately most of the names and dates are illegible, due to the fact that everyone wrote his name across someone else's. Plus some are faded and others are covered with a thin film of flowstone. But there are enough discernable dates present to establish that the cuts were made sometime in (or before) the 1860s. A couple of dates are very clear, such as

Figure 14.23. Erica next to large stalagmites in Morrill Cave. Photo by Bob Biddix, March 22, 2008.

obvious trouble of making all these cuts, but they left virtually all of the cut-out pieces lying around; practically none of the cut-out pieces were taken out of the cave. This is one mystery that will be difficult to clear up.[5]

Recent Discoveries In Morrill Cave

On an April 22, 1978, trip into Morrill Cave, caver Dan McDowell noted a rare cave salamander in the cave. He describes his find as follows:

> The stream in the lower level is from a few inches to 3 feet deep. We saw two species of salamanders, one of which appeared to be *Gyrinophilus porphyriticus*. It was rather large, 7 to 8 inches long, thick bodied, and a pale pinkish color. It had the feathery larval gills that are indicative of a neotenic state. We also saw a rather large crayfish that was very light in color, possibly troglobitic.[6]

Caver Laurie Adams, along with other cavers, surveyed an additional 5,000 feet of

"J.W. Dunn – July 11, 1877" and "Greene – 1867." Just beyond the tombstone area is the Signature Room, which contains hundreds of old names and dates. There is even a party of 13 explorers, half of which were women, dated Aug. 17, 1871. There is a signature by R.O. Worley, but unfortunately there is no corresponding date. There is even an obscure date: M. Jones – 1811 or 1812, which would be the oldest known date in the cave, however it is too muddled and unclear to be certain about it. Not only did someone go to the

5 Laurie Adams, op. cit.

6 Dan McDowell, "Morrell Cave, Sullivan County, Tennessee," *The Windy City Speleonews*, vol 18 (1978), no. 3, pp 60–61.

passage in Morrill Cave in the 1980s. Much of this additional footage was found in a long extension to the Railroad Passage. Peccary tracks were found in a remote section of this passage, and a bone deposit was found nearby with peccary, elk, and a number of unidentified bones.

This discovery is especially significant, since peccaries have been extinct in Tennessee for at least 10,000 to 12,000 years. Members of the peccary family still survive in Central and South America in much warmer climate than we have in Tennessee at the present time. Therefore, it is likely that these extinct peccaries lived here in Tennessee prior to the last Ice Age.

Figure 14.24. Erica Sughrue in the stream passage in Morrill Cave. Photo by Bob Biddix, March 22, 2008.

Morrill Cave as a State Natural Area

The Tennessee State Natural Area program began in 1973, with the designation of Virgin Falls as the first State Natural Area. Natural Areas generally are areas of unusual scenic beauty, worth preservation, but smaller in scale than State Parks. Caver Richard Geer gives the following information about the status of Morrill Cave as a State Natural Area in 1978:

> Talk of state acquisition has prevailed for perhaps 15 years. At one time east Tennessee's affiliated caving community tried to persuade the state to purchase the cave as a means of preserving it. This movement seems to have ended in frustration. But finally, the 1974–1975 State Legislature appropriated $30,000 with which to purchase a 45.6-acre tract which contained the 6-acre Morrell Sink and the cave's two entrances.
>
> About this time, condemnation proceedings were in progress to obtain a tract of slightly less than 5 acres at Sycamore Shoals near Elizabethton,

> Tennessee which is near by. The court awarded the owners of that land an exorbitant sum amounting to something like ten times the negotiated price of Morrell Cave, and the owners of the Morrell Cave property promptly demanded a substantially larger sum of money which was beyond the state's willingness or ability to pay. Negotiations for purchase of Morrell Cave collapsed.[7]

The state never purchased Morrill Cave and today (2008) it is still in private ownership.

Morrill Cave Today

Morrill Cave is currently what is referred to as a Tennessee "State Natural Area." As such, it is open to the public for self-guided tours. However, the cave is still privately owned. According to the Tennessee State Natural Areas Web site:

> The site is privately owned. Visitors must follow the directions provided below to the landowner's house. Upon arriving, go to the front porch of the landowner's house where the sign-in sheet and the cave gate keys are kept. Sign in and obtain a key to the cave gate. Lock the gate behind you after entering the cave and secure the key where it will not be lost. Lock the gate after you leave the cave and return the key to the porch before leaving the site. You may go into the cave even if the landowner is not home.

For more information on Morrill Cave, you may contact the Tennessee Department of Environment and Conservation at the following address:

Division of Natural Areas
7th Floor, L&C Annex
401 Church St
Nashville, TN 37243-0447

The telephone number for the Division of Natural Areas is:

615-532-0431

The Web site for the Division of Natural Areas is:

http://www.state.tn.us/environment/na/natareas/

This is a large, wild cave. Persons not experienced in cave exploration are not advised to enter this cave. It is surprisingly easy to get lost in a cave that is 5 miles long. Information on how to learn safe caving techniques is given in Appendix A.

Directions

From Elizabethton take U.S. Highway 19E/State Route 37 north toward Bunker Hill and Bluff City. Turn right on Chinquapin Grove Road. Stay on this road and travel north toward the South Fork of the Holston River. Turn left, heading west on Dry Branch Road, paralleling the South Fork of the Holston River. Turn left on Timber Ridge Road. Go about one mile. Entrance to the property is on the left.

Side Trips

From Elizabethton, drive southeast on U.S. Highway 19E for approximately 15 miles. Turn right (south) onto Tennessee Highway 143. After approximately 7 miles you will come to Roan Mountain State Park, which is nestled at the base of Roan Mountain. Continue on Highway 143 for approximately 7 miles to the crest of Roan Mountain and the Tennessee–North Carolina border. The crest of the Appalachian Mountains in this area is dominated by rhododendron balds, which are areas of grass intermixed with native rhododendron bushes. When these bushes bloom in late June, you will be rewarded by hundreds of acres of pink flowers. Many people call the state park in advance for updates on the timing of the display. The display will vary by several weeks from year-to-year, depending on the weather.

---

7 Richard P. Geer, "Morrell Cave," *The Michiana Caver*, vol 5 (1978), no. 3, pp 11, 15–17.

Even if you cannot time your visit to coincide with the blooming season, a trip to the top of Roan Mountain is not to be missed. The Appalachian Trail passes along the crest of the mountains here and offers wonderful opportunities for day hikes. The views are absolutely spectacular at any time of the year. Roan Mountain has become a popular destination for cross-country skiers in winter months.

Appendix A

Do You Want To Be A Caver ?

Just the fact that you are reading this book and are interested in the caves of Knoxville and the Great Smoky Mountains is an indication that you may want to become a cave explorer. Exploring caves can be a lot of fun; but, unfortunately, it can also be dangerous if you are not properly trained and equipped. There is a national organization, the National Speleological Society (NSS) that can provide you with information on local chapters in your area and can put you in contact with local cave explorers. They can be easily reached by e-mail on the World Wide Web at:

nss@caves.org

If you are not a computer user, you can write them at the following address:

The National Speleological Society
2813 Cave Ave
Huntsville AL 35810-4431

The NSS can also be contacted by telephone at:

256-852-1300

The NSS has a bookstore that sells books on caving and caving techniques. Several suppliers of caving equipment advertise on their Web site and in their publications. Their Web site is located at:

http://www.caves.org

If you do decide to explore caves, please remember the NSS' motto:

TAKE NOTHING BUT PICTURES
LEAVE NOTHING BUT FOOTPRINTS
KILL NOTHING BUT TIME

Knoxville and the Great Smoky Mountains are located in one of the cave rich areas of the United States. The NSS has active chapters (grottos) in east Tennessee that you may wish to join, if you live in this area. Never go in a cave without proper training and proper equipment. Some commercial caves offer wild cave tours. These are an outstanding way to learn safe caving techniques from experts before you venture out into wild caves on your own.

Have fun and be safe.

Appendix B

Glossary

A

Acetylene – A flammable gas (C_2H_2) formed by combining carbide and water.

Alum – A mineral composed of potassium aluminum sulfate, $KAl(SO_4)_2{-}12H_2O$.

Anthodite – A cave formation consisting of radiating or branching crystals of calcite or aragonite.

Aragonite – A mineral composed of calcium carbonate ($CaCO_3$), the same as calcite, but forming orthorhombic crystals instead of hexagonal crystals as calcite does.

Archaeologist – A scientist who studies human history and prehistory.

Archaeology – The science the studies the material remains of past human life.

Ascenders – Mechanical clamping devices for ropes with a ratcheting action that allows them to be raised easily, but to lock when weight is applied. They are placed on the rope and then attached to the body and feet with slings.

B

Bacon Rind – A thin, banded flowstone formation that resembles a strip of bacon.

Bat – A nocturnal flying mammal of many different species. They frequently spend the daylight hours inside caves and some hibernate inside caves during the winter.

Bedding Plane – The contact between two layers of sedimentary rocks.

Belly Crawl – A crawlway so low that it can only be negotiated flat on your stomach.

Biology – The science that studies the living organisms of the Earth.

Breakdown – A pile of rocks in a cave room or passage, formed by the collapse of the cave roof.

Breakdown Block – An individual piece of rock in a pile of breakdown.

Breakdown Slope – A sloping surface inside a cave, composed of or covered by breakdown.

C

Cable Ladder – A flexible ladder consisting of two small steel cables with small, aluminum rungs. These are available commercially in 10-meter sections. They can be rolled into a small bundle for easy transport through a cave. Two or more sections can be connected to make longer lengths.

Calcite – A mineral composed of calcium carbonate ($CaCO_3$), which is the main component of limestone. Calcite is also the main component of most cave formations, such as stalactites and stalagmites.

Canyon – A cave passage that is several times as high as it is wide.

Carabiner – A metal snap link used in technical climbing, especially to rig ropes.

Carbide – A man-made chemical, calcium carbide (CaC_2). Reacts with water to form acetylene gas. $CaC_2 + H_2O = C_2H_2 + CaO$.

Carbide Lamp – A lamp, usually manufactured from brass, which combines carbide and water to produce acetylene gas, which burns with a very bright flame.

Cave – A naturally occurring cavity in the Earth's crust, which is large enough to permit human exploration.

Cave Cricket – A species of cricket specifically adapted for life in caves. During mild weather some species of cave crickets leave their cave at night to forage for food on the Earth's surface.

Cave System – A group of caves connected by exploration. Frequently used to describe a large cave with several entrances.

Caver – A person who explores caves. This is the term cave explorers use to describe themselves.

Cavern – A term used to denote a very large or extensive cave.

Chamber – A large room-like portion of a cave.

Charcoal – A black organic substance formed by burning wood in an oxygen poor environment, resulting in incomplete combustion. Charcoal is composed primarily of organic carbon.

Chert – A microscopically grained quartz mineral (SiO_2) of various colors that forms beds, lenses, and nodules in limestone. Frequently used by prehistoric peoples to make stone tools.

Chimney – n. A passage narrow enough that you can climb up or down by placing your hands and feet on the opposite walls.

v. To climb up or down by placing your hands and feet on the opposite walls of a cave passage. The back is also occasionally used as a point of contact.

Clay – A very small, microscopic grain of sediment or dirt, usually composed of hydrated aluminum silicates.

Climb – n. A vertical section of a cave that requires climbing skills to ascend or to descend.

v. To ascend or descend a vertical section of a cave.

Column – A cave formation that extends from the ceiling to the floor.

Commercial Cave – A cave that has all or part of its passages developed with walkways, steps, and lights and is open to the public for guided tours. A show cave.

Connection – A naturally occurring passage which, when discovered, connects two caves which were previously considered separate caves.

Conservation – The preservation and wise use of our natural resources.

Copperas – A green, hydrated ferrous sulfate, $FeSO_4{-}7H_2O$, used primarily in making inks and pigments.

Crawl – n. A horizontal section of cave that requires crawling on your hands and knees or stomach to negotiate, due to a very low ceiling height.

v. To move on your hands and knees or your stomach.

Crawlway – A cave passage that is so low that cave explorers must crawl on their hands and knees or stomachs.

Crystal – The solid form of a mineral, which displays symmetrically arranged plane surfaces. This external shape is determined by the internal atomic structure.

Curtain – A thin, wide, frequently wavy speleothem hanging from the ceiling or wall and resembling a curtain. Also called a drapery.

D

Dip Slope – A geological term that refers to where the slope of the land surface is approximately the same as the dip of the underlying rocks.

Dome – A vertical passage extending upward from the ceiling of a cave passage or room.

Domepit – A vertical passage extending both upward and downward in a cave passage or room.

Drapery – A thin, frequently convoluted, sheet-like flowstone formation.

Dripstone – Cave formations composed of calcite or aragonite, formed when dripping water deposits dissolved limestone.

Drop – A pit or a cliff that must be rigged with rope or a ladder to descend.

E

Entrance – A point where a cave opens to the surface and is large enough to permit human passage.

Epsom Salt – A hydrated magnesium sulfate, $MgSO_4{-}7H_2O$, used as a laxative.

Escarpment – A steep slope, usually with cliffs, that separates two relatively level areas of different elevations.

F

Fill – Secondary deposits of sediments in a cave passage, usually consisting of clay, silt, sand, gravel, or a combination of two or more of these.

Flint – A lustrous, microscopically grained quartz mineral (SiO_2), usually gray or black, which forms beds, lenses, and nodules in limestone.

Flowstone – A cave formation composed of calcite or aragonite, formed when flowing water deposits dissolved limestone.

Formation – A secondary mineral feature in a cave, such as stalactites, stalagmites, columns, helictites, soda straws, shields, bacon rind, and the like.

Fossil – The remains or traces of animals and plants that lived in prehistoric times.

G

Geologist – A scientist who specializes in the study of geology.

Geology – The science that studies the structure, history, and origin of the Earth.

Grotto – A term used to describe a single cave room. Also the term used for a chapter of the National Speleological Society.

Ground Water – Any water below the surface of the ground.

Guano – The accumulated excrement of cave-dwelling animals, especially bats.

Gypsum – A mineral composed of hydrous calcium sulphate ($CaSO_4{-}H_2O$), which forms colorless to white crusts and crystals in a variety of forms.

Gypsum Crust – A layer of fibrous or crystalline gypsum that can occur on cave walls and ceilings.

Gypsum Flower – A flower-like group of curved gypsum crystals.

H

Hand Hold – A projection, depression, or other irregularity in the rock that can be used by a climber to support himself with his hand.

Hard Hat – A protective helmet worn by cave explorers to protect their heads from bumping the ceiling in low passages, from falling rocks dislodged by cavers climbing above them, and in case of a fall.

Helictite – A cave formation that grows in bizarre, twisting shapes apparently without regard to gravity.

Hydrology – The scientific study of water and its distribution and movement.

J

Joint – A fracture in sedimentary rocks, usually perpendicular to the bedding planes.

K

Karst – A characteristic topography found in areas of limestone bedrock, characterized by caves, sinking streams, sinkholes, and other features.

L

Leach – To remove minerals by dissolving them in water.

Lead – An unexplored passage in a cave.

Ledge – A horizontal projection along the side of a wall.

Limestone – A sedimentary rock composed primarily of calcite.

Lower Level – A series of cave passages developed at a lower elevation than the rest of the cave.

M

Magnesia – Magnesium oxide, MgO, used as an antacid and a mild laxative.

Map – n. A two-dimensional representation of a cave, showing its horizontal extent that may include cross-sectional representation to also show the vertical extent.

v. To gather the data necessary to draw a map.

N

National Speleological Society – An organization of cave explorers and speleologists, founded in 1941, to promote the exploration, preservation, and scientific study of caves. NSS.

Native Americans – This is the "politically correct" term for the people who lived in North America before the arrival of Europeans in 1492. These people generally prefer to be known as American Indians or by their tribal name.

Nitrates – Chemical compounds that contain the NO_3^- ion. Nitrate minerals are highly soluble in water.

Nitrogen – A colorless, odorless gas that makes up 78% of the Earth's atmosphere.

NSS – The National Speleological Society.

O

Onyx – A term used to describe lustrous, sometimes layered, flowstone.

P

Paleontologist – A scientist who studies fossils to learn about the history of life on Earth.

Paleontology – The science that studies the history of life on Earth.

Palette – A disk-shaped flowstone formation, usually growing from the wall of a cave. Also referred to as a shield.

Passage – A horizontal section of a cave.

Phreatic – Refers to water below the water table. Also, cave passages that form beneath the water table.

Pit – A vertical passage extending downward from the floor of a cave passage or room.

Pitch – A vertical section of cave that requires difficult climbing or rope to negotiate.

Popcorn – An irregular flowstone or gypsum formation with many rounded, knob-like surfaces.

Prussik – To ascend a rope by any of a variety of methods.

Prussik Knot – A knot that will slide easily when there is no weight on it, but that will hold securely when weight is placed on it. By using two or three Prussik knots, a climber can ascend a standing rope.

R

Rappel – To descend a rope by any of a variety of methods.

Rig – To secure ropes or cable ladders so that a drop can be safely descended.

Rimstone – A raised flowstone edge surrounding a pool of water.

Rimstone Dam – A naturally formed flowstone dam occurring at the downstream edge of a cave pool.

Rimstone Pool – A pool of water surrounded by a raised edge of flowstone.

Rock – The solid material that forms the Earth's crust.

Rock Shelter – A rock overhang that resembles a cave entrance but does not open into an actual cave.

Room – A room-shaped opening in a cave.

S

Saltpeter – A naturally occurring mineral, calcium nitrate, $Ca(NO_3)_2$, used in the manufacture of gunpowder.

Saltpeter Cave – A cave with enough saltpeter in the dirt to have been mined commercially at some time in the past.

Saltpeter Vat – A vat constructed to leach saltpeter from cave dirt.

Saltpetre – A spelling of saltpeter that was widely used in the 1800s.

Sandstone – A sedimentary rock composed primarily of cemented sand grains.

Scallops – Shallow, rounded pockets in a cave wall, formed when water was flowing in the cave passage.

Sedimentary Rock – A rock formed by the deposit of water-borne and/or wind-borne particles or by the accumulation of chemically or organically precipitated materials.

Shale – A sedimentary rock composed primarily of cemented clay particles.

Shield – A disk-shaped flowstone formation, usually growing from the wall of a cave. Also referred to as a palette.

Show Cave – A commercial cave.

Side Passage – A cave passage that leads off of another cave passage.

Silt – A small grain of sediment or dirt, approximately the size of a particle of flour.

Sink – n. An abbreviated form of the term sinkhole.

v. For a surface stream to disappear underground.

Sinkhole – A surface depression in an area of limestone rock formed either by solution as water drains underground or by the collapse of a cave passage or room.

Sinkhole Plain – An area of relatively flat topography with numerous sinkholes.

Sinking Stream – A surface stream that abruptly disappears and goes underground into a cave.

Siphon – A section of water-filled cave passage connecting two sections of cave passage with air is frequently referred to as a siphon. However, true siphons rarely, if ever, occur in caves. The more proper name for these water-filled sections of cave passage is a *sump*.

Skylight – A natural opening to the surface in the ceiling of a cave passage or room.

Soda Straw – A thin, hollow stalactite, similar in size and shape to a drinking straw.

Soluble – Capable of being dissolved, usually referring to water.

Speleogenesis – The origin and development of caves.

Speleologist – A scientist who studies some aspect of caves, such as their biology, geology, meteorology, and the like.

Speleology – The scientific study of caves, their origins, and their features.

Speleothem – A secondary mineral deposit in a cave.

Spelunker – A person who explores caves. Cave explorers rarely use this term to describe themselves, preferring instead to call themselves cavers.

Spring – A place where water naturally flows out from underground onto the Earth's surface.

Squeeze – n. A section of cave passage so small that it is just barely possible to fit through.

v. To move through a very small section of cave passage.

Squeezeway – A passageway so small and tight that the explorers must squeeze to get through it.

Stalactite – An icicle-shaped cave formation that hangs down from the ceiling.

Stalagmite – A conical-shaped cave formation that grows upward from the floor.

Stoopway – A cave passage that is too low to stand up completely, but not as low as a crawlway.

Sulfur – A yellow, non-metallic element, used in the manufacture of gunpowder.

Sump – A section of cave passage that is completely filled with water.

Survey – n: A map prepared by using an accurate compass, a measuring tape, a clinometer, and sketches.

v: To prepare an accurate survey or map.

Survey Station – A point in the cave from which compass direction, inclination, and distance are measured to the next point.

Surveyed – A cave or cave passages that have been accurately mapped.

Suunto – A brand name of a high-quality surveying compass.

T

Terminal Breakdown – A breakdown that completely blocks a cave passage.

Topographic Map – A map that represents the surface of the Earth by using lines to indicate points of equal elevation at selected intervals. These lines are referred to as contour lines.

Totem Pole – A tall, cylindrical stalagmite.

Traverse – n. A difficult ledge or section of wall that must be crossed to continue exploring a cave.

v. To move across a difficult ledge or to move laterally across a wall.

U

Unsurveyed – A cave or cave passages that have not been accurately mapped.

Upper Level – A series of cave passages developed at a higher level than the rest of the cave.

V

Vadose – Refers to cave passages that form above the water table.

Vandal – A person who vandalizes a cave.

Vandalism – Altering a cave, either accidentally or intentionally, by breaking formations, leaving trash, writing on walls, or otherwise changing the original appearance of the cave.

Vat – A container built to hold cave dirt to allow water to percolate through the dirt and remove the nitrate minerals.

Virgin Passage – Cave passage that has not been entered previously.

W

Walking Passage – A cave passage large enough to walk without stooping.

Water Table – The horizontal plane beneath which all open spaces in the Earth are filled with water.

Wild Cave – A cave that has not been developed for guided tours. Also, any portion of a commercial cave that has not been developed for guided tours.

Appendix C

Suggested Additional Reading

If you enjoyed this book you may wish to read other books on caves and cave exploration. Here are a few suggestions for books that are still in print and readily available at the time that this book went to the printers:

Big Bone Cave, Larry E. Matthews, 2006, National Speleological Society, Huntsville, Alabama, 220 pages.

The fascinating story of America's largest saltpeter mine, located 50 miles north of Chattanooga. Also the site of the discovery of giant ground sloth skeletons, an ice-age jaguar skeleton, and several American Indian mummies.

Cave Minerals of the World, by Carol Hill and Paolo Forti, 1997, National Speleological Society, Huntsville, Alabama, 463 pages.

A comprehensive guide to the minerals found in caves and the speleothems that they form. Includes many beautiful color photographs.

Caves of Chattanooga, Larry E. Matthews, 2007, National Speleological Society, Huntsville, Alabama, 192 pages.

The complete history of nine caves in the greater Chattanooga area that either currently are commercial caves or have been commercial caves in the past. This book is the perfect companion to this book, *Caves of Knoxville and the Great Smoky Mountains*.

Caving Basics, Third Edition, Editor G. Thomas Rea, 1992. National Speleological Society, Huntsville, Alabama, 187 pages.

This book is an excellent introduction for new cavers who want to master the basic skills.

Cumberland Caverns, Larry E. Matthews, 2005, Greyhound Press, Cloverdale, Indiana, 188 pages.

The complete history of the exploration of Tennessee's largest cave, located 70 miles northwest of Chattanooga and open to the public.

Dunbar Cave: The Showplace of the South, Larry E. Matthews, 2005, National Speleological Society, Huntsville, Alabama, 145 pages.

The exciting story of the exploration of this cave from 8,000 years ago to the present. This cave has a long, rich history and was the center of Clarksville, Tennessee's, social life for nearly 100 years.

On Call, Edited by John Hempel and Annette Fregeau-Conover, National Speleological Society, Huntsville, Alabama, 384 pages.

This is the most comprehensive book on cave rescue ever written.

On Rope II, by Allen Padgett and Bruce Smith, National Speleological Society, Huntsville, Alabama.

This is the most thorough guide available for rope techniques. These techniques are used for cave exploration, mountaineering, and rescue.

Speleology – Caves and the Cave Environment, George W. Moore and Nicholas Sullivan.

This is an easy-to-understand introduction to the science of caves, including their biology, geology, and how they form.

The Hidden World of Caves - A Children's Guide to the Underground Wilderness, Ronal C. Kerbo.

An excellent introduction to caves for younger readers.

There are several mail order firms that carry an extensive variety of books and magazines on caves. Most of these sources advertise in the *NSS News*. The National Speleological Society Bookstore carries a large number of caving books. A current list of what is available can be obtained by writing them at this address:

NSS Bookstore
2813 Cave Ave
Huntsville AL 35810-4431

Telephone: 256-852-1300

You can also find the NSS Bookstore on the World Wide Web at:

http://www.nssbookstore.org

There are hundreds of books on caves that have been printed in the English language. You will have a wonderful choice to choose from at the above sites.

Index

C

D

I

J

K

L

M

R

S

T